ALPINES

An essential guide

ALPINES
An essential guide

Michael Mitchell

THE CROWOOD PRESS

First published in 2011 by
The Crowood Press Ltd
Ramsbury, Marlborough
Wiltshire SN8 2HR

enquiries@crowood.com

www.crowood.com

Paperback edition 2022

British Library Cataloguing-in-Publication Data
A catalogue record for this book is available from the British Library.

ISBN 978 0 7198 4085 2

Disclaimer
The author and publisher do not accept any responsibility in any manner whatsoever for any error or omission, or any loss, damage, injury, adverse outcome, or liability of any kind incurred as a result of the use of any of the information contained in this book, or reliance upon it.

Frontispiece: *Lewisia cotyledon* hybrids.

Cover design by Maggie Mellett

Typeset by Servis Filmsetting Ltd, Stockport, Cheshire

Printed and bound in India by Parksons Graphics Pvt Ltd

Contents

Preface

A north-facing barren hillside, 900ft up in the Yorkshire Pennines, would not be an ideal place to choose to set up an alpine garden and nursery. But it wasn't choice, rather necessity and circumstance. Having just left Askham Bryan horticultural college near York in 1986, I was keen to start my own nursery and a field below my parents' garden came up for sale... so that was the beginnings of Slack Top Alpine Nursery.

Several years of hard slog followed, turning rough hillside pasture into a suitable growing site: covering an acre of land with gravel using a wheelbarrow, shovel and rake after first having removed hollows and hillocks with the shovel and rake (finances were non-existent with which to hire equipment or labour); putting up home-made greenhouses and cold frames; planting hundreds of trees for shelter; erecting polytunnels, and carrying and laying about a mile of concrete paving slabs for nursery pathways. Not a way I would recommend now with the benefit of hindsight, but as an enthusiastic twenty-one-year-old starting out it was the only option. I believe the phrase is 'character forming'!

As a prospective location for a nursery and garden site, an exposed north facing hillside (sometimes known as a 'slack') doesn't seem ideal, but actually the difficult climate is probably one of the nursery's main benefits. The high rainfall, cool temperatures and short growing season weeds out plants on the edge of hardiness resulting in really tough plants.

Alpine Garden at Slack Top, Hebden Bridge, UK.

The climate at Slack Top is a challenging one with high rainfall throughout the year, often at a time when it is least required, and dull damp weather never very far away. Yet despite such difficulties alpines thrive even here, demonstrating what a tough and adaptable group of plants they are if their few simple requirements are met.

Alpine plants are my passion because they are so varied and exquisitely beautiful. I have selected some of my personal favourites to feature in this book – hardy plants that perform really well in our Pennine garden and put on an unrivalled display of flowers, foliage and form the whole year round. As many of our nursery customers have exclaimed over the years 'if it'll grow here, it'll grow anywhere'!

OPPOSITE PAGE:
Sempervivum arachnoideum has both beautiful flowers and foliage.

1 An introduction to alpines

Mixed alpine containers give maximum impact in a small space.

WHY GROW ALPINE PLANTS?

Beauty in miniature

There are many good reasons for growing alpines but surely one of the best is to enjoy and marvel at their exquisite beauty and great diversity. With hundreds of different alpines widely available and thousands more grown by specialist nurseries there are plenty to choose from. With some careful selection it is possible to have a lovely display of colour, form and foliage interest during every month of the year, with the bonus that the majority of alpine plants are perennial and will repeat their performance for many years to come.

OPPOSITE PAGE:
A striking red form of *Pulsatilla vulgaris* 'Papageno'.

Small spaces

Nowadays, gardens are victims of the increasing demand for building space, with the result that new homes generally have a smaller plot of land on which to accommodate not only a garden, but also all the other demands of modern living, such as parking, garden shed, patio, and so on. In many instances, once these facilities have been catered for there is perhaps only enough space left for one type of garden feature. Unlike an herbaceous border or shrubbery an alpine bed can be made in the smallest of spaces with several species growing together in an area as small as 1m^2. A garden feature can easily be created to showcase the plants – whether a raised bed made from timber, brick, stone or perhaps concrete painted to complement the plants and surroundings, or several containers grouped together on a low wall surrounding an outdoor eating area. Even a rockery can be as small or large as the space dictates, with larger constructions allowing the indulgence of rock outcrops and ravines or even a cascade of water into a pool.

Tough survivors

By virtue of their natural habitats alpines are very hardy plants, well able to survive the harsh conditions that mountains inflict on them. Anyone who has visited the mountains in winter can appreciate that plants from these regions are unlikely to be troubled by any of the cold experienced in domestic gardens at lower altitudes, and even in summer temperatures in the mountains can exceed 30°C (86°F). A successful alpine garden can therefore be made in many parts of the world. Gardens that are exposed to strong winds are not a problem either, as even those described as 'wind tunnels' are unlikely to be any windier than a mountainside.

The toughness and adaptability of alpine plants makes many of them very suitable for anyone new to gardening, with some species practically thriving on neglect, which can be very encouraging for children and beginners alike. At the other end of the spectrum some species are very difficult to grow even for experts and so there is never any likelihood of the enthusiast running out of challenges.

Undemanding

Most alpines require only a well-drained soil, and a site not overhung by trees or in deep shade. Because of their small stature and adaptations to an alpine climate many of them have minimal watering requirements when compared to shrubs or herbaceous plants. Maintenance in a garden setting is minimal, making these plants suitable for anyone with only a small amount of spare time, unless they are being grown in pots in an alpine house, in which case regular observation is required for watering, shading and re-potting. With most being perennial they are a relatively permanent planting outdoors. The majority have a natural lifespan of several if not ten or twenty years if all goes well, needing perhaps only an annual trim to keep them tidy or maybe dividing every few years in order to maintain vigour.

Alpines are also eminently suitable for young or old alike who may find it difficult to physically work on a ground level garden; by having raised beds, containers or an alpine house, plants are brought to a height allowing easier planting, tending and of course admiration!

Remember also that no garden is too small to grow alpines. Even a simple window box provides suitable space for a collection of plants – with sun lovers such as *Sempervivum*, *Sedum* or *Lewisia cotyledon* for a southerly aspect or *Ramonda myconi* and many of the primulas for a more northerly one offering cooler, shadier conditions.

ALPINE PLANT CHARACTERISTICS AND ADAPTATIONS

The word 'alpine' in relation to plants means any that are found growing above the tree line in mountainous areas. An alpine plant, from the gardener's perspective, is any plant suitable for growing on some form of rock garden, be it a raised bed, scree, sink, container or rockery. What this means in practice is a plant that is hardy, relatively small and has its origins in mountainous areas of the world such as the Alps of Europe, the Rocky Mountains of North America, the Himalayas or the mountains of New Zealand. Habitats in these areas include high peaks and cliffs, scree slopes and high meadows that can be orientated in any of the four aspects, thereby receiving varying amounts of sun, shade, wind, rain and temperature.

Such varied conditions have resulted in many alpines developing adaptations in order to help them survive. Many of the same adaptations can be seen among different species, for example *Helichrysum milfordiae*, *Sempervivum arachnoideum*, and *Leucogenes leontopodium*: all have silver or grey foliage, with the leaves covered in hairs which help to insulate them from the cold as well as the heat of the sun, and conserve moisture on arid sites. Others, such as *Androsace*, *Armeria* and *Vitaliana primuliflora* grow close to the ground as dense cushions or mats of tiny leaves, giving them as small a surface area as possible to reduce exposure to the drying and

A characteristic alpine meadow in the Aosta valley, Italian Alps.

chilling effects of high winds, often on sites so exposed that taller plants would simply be blown away or broken down. Species growing among the loose stones of screes often have flexible creeping stems or runners, allowing them to survive any movement among the rocks, for example *Linaria alpina* and certain *Campanula* species. A characteristic common to a great many alpines is one of brilliantly coloured flowers for attracting insect pollinators from across what can be quite barren mountain landscapes.

Contrary to popular belief, alpines make a considerable amount of root considering their physical size. Plants growing in crevices among rocks will quite likely have to delve deep in order to access both moisture and nutrients, with those growing among scree facing a similar situation.

Methods of seed dispersal are wide ranging with crevice and cliff dwellers like *Saxifraga longifolia* and *Physoplexis comosa* producing hundreds of very small seeds that simply fall from the plant to find by chance suitable places for germination. The seeds of *Pulsatilla* species have bristly appendages that attach themselves to the fur of animals or are blown away on winds to reach new sites. Those of *Erigeron* are like little parachutes that catch the wind and may travel hundreds of metres before landing and being blown around until they become lodged somewhere.

Germination characteristics of alpine seed are especially interesting, with many species requiring a prolonged period of cold temperatures, often with alternate freezing and thawing before germination takes place once warmer weather arrives in the spring. This is probably to give seedlings the maximum length of growing time in the short seasons encountered in mountainous areas.

THE BINOMIAL SYSTEM OF PLANT CLASSIFICATION

Every alpine plant, just like any living or once living thing, is given a Latin name to enable anyone anywhere in the world to identify it. A Latin name is universally understood whereas common names can differ across the globe. The first part of a Latin plant name is made up of the genus. Plants possessing certain specific botanical characteristics are placed within a genus but in order to differentiate plants within a genus they are each then given a species name. For example, in *Primula vulgaris*, *Primula* is the genus and *vulgaris* is the species within that genus. All plants have both genus and species after which there may be further separation into subspecies, variety and form. By this system every known plant is given a unique name. Plants of differing genera are also grouped into families, for example the family Primulaceae contains a number of genera e.g. *Primula*, *Androsace* and *Cyclamen*.

Apart from being a basis for researching a plant, Latin names can also provide fascinating clues about a plant's physical characteristics or natural habitat; perhaps sometimes commemorating the person who first found it or the country in which it was discovered.

What's in a name?

Here are a few Latin alpine plant names and what they reveal:
Sempervivum – translated from Latin as 'ever living', and probably so named because of the plant's toughness.
Sanguinaria – refers to the blood-red colour of juices contained in the plants' rhizomes. Sanguineus in Latin means 'of blood'.
Hepatica – having liver-shaped leaves. The Latin word hepatica means 'plant having liver shaped parts, or one used to treat liver diseases'.
juniperifolia – means 'with leaves like a juniper'.
verna – refers to a plant's springtime flowering, from Latin vernalis, 'of the spring'.
alpinus – grows on high mountains.
darwinii – given to a plant named in honour of its discoverer Charles Darwin.

2 Traditional settings

The earliest methods of cultivating alpine plants generally involved them being planted in a naturalistic man-made setting in what we think of as a rock garden or perhaps a scree, where many different species could be grown together in a well-drained soil. Simply put, rockwork (sometimes homemade from a sand, gravel and cement mixture) was used to create an outcrop and the alpines planted there. The use of stone sinks or troughs in which to house a collection of alpines has also been widely practised for many years. These traditional locations for cultivation along with raised beds and alpine houses are still very popular and ideally suited to today's gardens.

Colourful corner of a rock garden.

THE ROCK GARDEN

Probably the most recognized and popular way of growing alpine plants is on what is usually called a rock garden or rockery, generally built on a sloping site and incorporating rockwork of some sort. Carefully constructed, it can provide suitable conditions for a very wide range of alpine plants. A rockery can be of any size and therefore is suitable for both small and large garden spaces. Almost any garden can be transformed by the addition of a rockery; lawns have their uses, but a miniature alpine landscape with exciting, beautiful and fascinating plants giving interest all year round, offers far more than a patch of grass.

OPPOSITE PAGE:
Stone troughs, raised beds and paving all provide planting opportunities for alpines.

The base of the rockery is generally where excess rainwater collects and eventually drains from. This area is ideal for the cultivation of plants requiring plenty of water and which can on occasion tolerate being submerged for short periods, at least for a few hours. Such plants could include bog primulas, perhaps most spectacularly represented in a group known as 'Harlow Carr Hybrids'. Above this area, a carefully contrived outcrop of rocks could be softened by mat-forming or trailing alpines – *Phlox*, *Aubrieta* and *Lithodora* to name a few from hundreds of possibilities. Higher still, among more rocks and with sharper drainage, conditions would be suitable for plants like the fleshy leaved *Lewisia cotyledon* with its gaudy flowers in pinks and oranges. If conditions were particularly suitable, then it may self sow and create a wonderful and naturalistic drift. Although a sunny aspect is generally desirable for a rock garden, any shady corners and crevices have plenty of desirable treasures to choose from. *Ramonda myconi* would be a beautiful option, growing

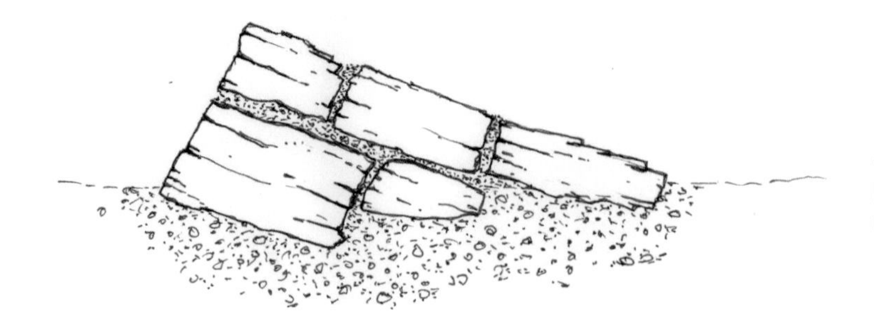

Cross section of a suggested layout showing backwards tilt of rockwork on a level site.

Cross section of rockwork on a sloping site.

naturally in the Pyrenees where it inhabits cool, shady rock faces. Delicate purple flowers belie its toughness and it can thrive in the British climate if given a well shaded spot.

Rockwork

A sloping garden makes it easier to build a naturalistic rock outcrop that appears to have been exposed by the erosion of surrounding soil, rather than having to build up rockwork from a level site. If you are unfamiliar with the appearance of a natural outcrop of rocks, then it may be worth taking some time to study a few locally as it will certainly help, when confronted by a pile of rocks weighing several tons, to know which to start with and where.

The drawings illustrate how rockwork can be arranged and there are a couple of key points that are essential when placing rocks. Firstly, they should have a slight backwards lean to them, as this serves to run rainwater back into the soil/compost, rather than creating a waterfall effect with a pool at the base! Also, natural outcrops rarely have rocks in isolation – in other words rocks 'join on' to one another, either side by side or one on top of another. Rocks that have been quarried will most likely have one or more 'faces' – that is, a flat side – and this helps with positioning. Keeping faces in the same plane, whether at right angles to the ground or to the top, will help to achieve a natural appearance. Rocks that have not been quarried are likely to be more rounded in nature and are more difficult to place, so unless a challenge is relished be sure to visit the quarry and specify the type of rocks you would like.

The stability of rocks is very important, as they can be utilized as stepping stones in order to be able to carry out maintenance work, or simply to get a closer look at the plants. To this end, remember to bury the bases at least a few inches and make sure the soil underneath is firm to ensure there is no possibility of them toppling over. As work progresses, remember to continue making sure that the compost is firmed around and in between the rocks, as any loose soil left at this stage may cause rocks to lean at an unwanted angle or give way underfoot if they are to be used as stepping stones. All this may seem like a lot to take in but once building starts, progress is soon achieved.

Rock types

The type of rock chosen to make a rock garden will have an effect on both its construction method and finished appearance. This is because a rock's strata tend to dictate the size and shape of the rock. Sandstones are some of the best rocks to use as they often have obvious strata lines that help with their placement and they are usually available in manageably sized pieces. When selecting rock at a quarry try to envisage how they might fit together in the garden. If the rocks don't seem to show a natural fit then it may be best to look elsewhere. Rounded boulders or those without an obvious 'face' can be very hard to build with.

The type of stone to use is probably best dictated by that which is locally available. Our nursery and display garden is located in the central Pennines, where millstone grit is the local rock and therefore sandstones are readily available from the numerous quarries in the area. Using local stone means a rock garden will sit more comfortably in its surroundings. Large rocks look best, kept in scale with the rest of the garden, but remember that they will have to be manoeuvred into position and it may be some distance from delivery point to the site of the rockery. A sack cart with pneumatic tyres can be used to transport up to 50kg (110lb) – and this is not a particularly large rock! – with a couple of strong people and a rope; larger ones can be pulled over grass on a sheet of strong polythene, or maybe machinery can be hired. I remember asking a quarry to deliver stones that could be fitted into a wheelbarrow. Maybe there was a misunderstanding, but what came out of the back of a skip were stones the size of, if not larger than, a wheelbarrow itself!

The quantity of rock required naturally depends on the type of construction intended, so any figures mentioned here are given as a rough guide. Rock itself varies quite considerably in weight so bear this in mind also. For a construction covering an area of approximately 10m^2 (11.9 sq yd^2), with rockwork covering 50 per cent, the rest being soil, a typical feature might require around 3–4 tons (3,000–3,500kg) of rock, in various sizes from say 10kg to 100kg (22–220lb). Anything larger becomes difficult to move, and could begin to look out of proportion in an area of 10m^2. Useful tools for construction are a digging spade, hand trowel, lump hammer for packing soil, and easily the most useful of all, a heavy crow-bar for levering rocks into place. A sturdy pair of gloves will help reduce the chances of scraped knuckles and squashed fingers, and it is sensible to wear a pair of stout boots, ideally with steel toecaps.

Compost/soil

The compost or improved soil used in the construction of a rock garden requires one particular property critical to the success of alpine plants, and that is drainage. The ground

A natural limestone outcrop in the Yorkshire Dales.

Linaria alpina on a natural area of scree.

directly beneath should ideally be well drained; that is, any puddles that form during heavy rain should drain away within half an hour or so. If this is not the case then it should either be possible to put in some simple land drains, or dig over the area and incorporate gravel and/or sharp sand; this may cure the problem. If the rock garden is on a slope then drainage problems are far less likely to be an issue. Water collecting at the base, as mentioned earlier, can be a bonus in that it could be used to create a bog garden in the form of a temporary soakaway. Ensuring well-drained compost for alpines at the construction stage can seem a laborious process, but the wide range of plants it will be possible to grow as a consequence will more than repay the time and effort. There is nothing more annoying than finding out that carefully chosen plants are likely to struggle in a new rock garden simply because not enough sharp sand or grit was mixed in with the soil!

The majority of alpines are happily accommodated in a soil that is slightly acid to neutral in pH, and most garden soils fall into this range. Fertility should ideally be not too high as this can lead to plants producing too much foliage at the expense of flowers. Also, in rich soils alpine plants can often grow out of character and be more susceptible to severely cold weather. In fortunate circumstances where garden soil is well drained then it may not be necessary to add anything to it, though if a sandy soil, then it is likely to be short of humus. This can be provided in the form of the contents of a well-decomposed compost heap, leaf mould or perhaps by purchasing a quantity of loam.

A good compost mixture that will provide the necessary requirements in which to cultivate a wide range of species comprises: 4 parts by volume of garden soil or loam, to 1 part by volume of sharp sand or fine grit. The grit should be not too large in size, say 3–4mm (0.25in). Its purpose is to prevent the soil constituent from sticking together, thereby allowing excess water to freely drain away. As a guide, try picking up a handful of moist, not wet, compost and squeeze it into a ball. If this ball cannot be made to break up with a modest prod, then it may require more grit. With regard to the soil content, a proportion would ideally be made up of leaf mould or some other form of humus, though this is not essential. As already hinted, the addition of fertilizers is unlikely to be necessary, though hoof and horn applied at the lower end of the manufacturer's application rate should help with any doubts as to fertility. As regards pH, or lime content, aim for a neutral to slightly acid compost as this will allow for the widest range of plants to be cultivated.

THE SCREE GARDEN

Naturally occurring scree can provide the most rapidly drained conditions for alpine plants, being composed of up to a metre (40in) depth, perhaps more, of angular pieces of frost shattered rock varying in size from less than a centimetre to several centimetres, all on what is often a very steep slope. Plants growing in such a habitat often have lengthy stems that become buried due to the shifting nature of the scree. Roots also can be extensive in order to access the moisture and soil deep below, which is generally low in fertility. These conditions are possible to replicate in the garden with a few adaptations, creating a site suitable for those alpines that particularly enjoy, or require, the sharpest of drainage. A scree also provides an almost ideal site for self-sown seed to germinate and develop thereby helping to create a more naturalistic appearance.

Site and construction

A sloping site is a good choice on which to make a scree garden. Where a slope is not present it may be possible to simply build up one end of the bed to provide a gradient. A scree-type bed can also be built on a flat site and though its drainage may not be as rapid as one on a slope, this should not be a problem. It can be very effective to incorporate a scree bed into a large rock garden, perhaps by placing it between two outcrops on a slope leading down to a more moist area. The aspect of the proposed site isn't too important, as all points of the compass are faced in nature, though ideally an open situation should be chosen, therefore receiving some sun and certainly avoiding overhanging trees.

Although construction methods can vary considerably, something approximating to the following will give excellent results. Whilst a gently sloping site is advantageous, bear in mind that too much of a gradient will result in the compost/gravel mix steadily slipping downhill! A gradient of between 1 in 5 and 1 in 15 is usually about right. It is essential that excess water can quickly drain away either at the bottom of the bed or at the lower end of the gradient. This can either be done with ground that is naturally well drained, or if not, it should be possible to dig a sump or lead the water away with a simple drain.

Firstly, any topsoil is dug out of the chosen area and placed to one side for later use, and the subsoil beneath (or at least the next 20cm (8in) or so) removed and discarded, to be replaced by a layer of 20–30cm (8–12in) of gravel/shingle. On top of this gravel layer some of the topsoil previously set aside can be utilized to make the scree mixture, assuming it is of good quality, by mixing with it a quantity of grit and sharp sand in the approximate proportions by volume of 2 parts topsoil, 2 parts grit, and 1 part sharp sand. This layer can be anything from 15cm to 30cm (6–12in) or so – it doesn't really matter too much. Finally, finish off with a 3–4cm top dressing of grit (see section on top dressing, page 40).

Although this suggested method of construction is unlike a real scree, the resulting conditions will allow the cultivation of many alpines that flourish with extra sharp drainage, as well as being enthusiastically colonized by those which do not. A few well-chosen rocks may be added to the final bed, acting as stepping-stones to allow easy access to the plants, as well as giving a more natural appearance.

A scree garden provides extra sharp drainage that suits many alpine plants.

THE ALPINE HOUSE

An alpine house may sound like a rather grand affair, which indeed it can be. However, in its simplest form it is merely a well-ventilated greenhouse that provides a suitable habitat for the alpine enthusiast to indulge his or her hobby without the need to dig or get wet! Joking aside, an alpine house does offer a comfortable form of gardening during inclement weather, as well as conveniently bringing what are often tiny plants up nearer to eye level where they can be more easily appreciated and tended. Pottering about amongst alpine treasures is a lovely way to brighten up a miserable day. Looking at it from the plant's perspective, an alpine house allows protection from extremes of weather, whether rain, heat or indeed anything else likely to result in their demise. Many alpines also have flowers that can be spoilt by the prolonged bad weather that often occurs during spring. Although any alpine plant may be grown under cover, an alpine house is

Colourful Alpine House display at RHS Garden Harlow Carr, Yorkshire.

most useful for the successful cultivation of those species that particularly need to be protected from excess rain, particularly over the winter months. Its protective nature also enables water to be withheld at any time which can be useful with many bulbous plants, such as the South African *Rhodohypoxis*, which are best kept dry in winter. An alpine house also allows the cultivation of plants in pots containing different composts, tailor-made to an individual plant. Alternatively an indoor landscaped rock garden can be created where all the plants are grown in the same type of compost.

Requirements

As an alpine house is really just a well-ventilated and unheated greenhouse, it is generally a simple matter to modify an existing structure. Excellent airflow is required in order to fulfil the following two functions: to avoid overheating due to the warming effect of the sun; and to discourage the establishment of certain fungal diseases. A door at either end of the structure would be an ideal start as this encourages a through draught, and if the door is a sliding type so much the better as these are less likely to blow about in the wind. Conventional hinged doors could simply have a catch to fasten them back. Opening vents at the ridge can also usually be fitted, and some of the side panes of glass could be removed and replaced with louvre-type vents. It may be that not all of these suggestions can be accommodated, but it should at least be possible to have doors which can be left open and remove some panes from along the sides. An air circulation fan could be another useful addition, especially on hot days.

Shading

Even with the best ventilation an alpine house is likely to become too hot for alpine plants without some form of shading. This is particularly relevant from approximately spring to late summer, during which time shading regularly has to be left on permanently. This is usually unavoidable if the owner is away at work during the day, and therefore not able to remove or apply shading according to whether the sun is shining or not.

Shading need only be placed on the sunny side of the greenhouse in order to prevent direct rays from the sun entering. Shading that can

The shading and ventilation on this alpine house can easily be altered as required.

be quickly removed or applied is preferable, and avoid using a whitewash liquid which, once applied, has to be left on for the whole season until it is wiped off at the end of summer. Plastic netting with a shade value of 40–60 per cent is readily available, not too expensive, and is easily fastened to the outside of a wooden greenhouse with hooks or something similar (fittings are available for an aluminium structure). Shading is far more effective when on the outside of the glass: placed on the inside, it can only protect from scorching, since the sun's heat has already entered. So by simply using a combination of ventilation and shading, excessive temperatures and stagnant conditions can be avoided; these are amongst the main causes of plant loss in the alpine house.

Clay or plastic?

Alpine plants can be successfully grown in either clay or plastic pots though each requires a slightly different treatment and indeed dictates to some extent the type of staging used to accommodate plants. Because clay pots are porous, they tend to dry out rapidly during the growing season. So unless the alpine house can be checked for watering at least once every day or so in warm weather, clay pots should be plunged in damp sharp sand up to their rim. This not only allows excess water to pass from the inside of the pot to the sand (useful in cases of overwatering), but also plants rarely dry out provided that some moisture is present in the sand. Indeed, watering can be safely carried out by simply keeping the sand plunge damp, thus allowing the porous nature of the clay pot to keep its contents moist. Using this system overwatering is unlikely to occur which is very useful, as until an individual plant's requirements are learnt, it can be tricky to get watering right. Pots plunged in sand also keep the roots cooler than those on an open bench, and the clay pot itself is given some insulation/ protection from being damaged during winter frosts.

Plastic pots have several advantages over clay – being easier to clean, cheaper to purchase and virtually unbreakable. They can, with careful management, be used to grow even the trickiest of alpines to perfection. Remember that unlike clay, a plastic pot is not porous and therefore the compost contained therein retains moisture for longer. This of course means more care needs to be taken to avoid overwatering; conversely, drying out is less likely. For this reason alpines grown in plastic pots are normally placed directly onto a wooden slatted bench and not plunged in sand. Standing pots on a layer of damp sand may seem like a good idea but in practice the plant's roots soon grow out of the base of the pot and into the sand; so much so that when it comes time to re-pot, much of the plant's root system is torn off in an effort to get it out of the pot.

Compost for pot culture/ potting

Suitable composts for use in pots can be made using a few readily available ingredients. The following quantities, measured by volume, give a good general potting medium:

- 2 parts sterilized loam
- 1 part crushed grit 3–4mm (0.25in). Perlite or vermiculite could be used as substitutes.
- 1 part leaf mould (or alternatively peat or a peat substitute)
- A small amount of base fertilizer at the lower end of the recommended application rate.

The addition of ground lime may also be required to achieve a pH of approximately 6.5. The mix can easily be altered, say, by adding more grit if a particular plant requires better drainage. Plants requiring acid or ericaceous compost will need the loam constituent of the above mix to be acidic and the ground lime can be left out.

If all this seems a bit complicated or storage space for ingredients is tight, then do not despair. There is an alternative to making your own, which is to use a ready-made loam-based compost known as John Innes. It comes in three strengths, with No. 2 being the most versatile for alpine plants. To every 3 parts of John Innes No. 2, add 1 part of grit (or a substitute). Ready-made ericaceous composts are also available. For a beginner this is often the best way to get a feel for compost, and in fact many keen amateurs base their composts around John Innes compost.

Leaf mould

Leaf mould is simply the decayed flaky remains of leaves and makes a wonderful dark crumbly material that can be used as a superior compost additive. It offers a great alternative to peat as it provides a valuable source of food as well as improving soil structure. Unfortunately leaf mould is not generally available to purchase so gardeners need to make their own. Fallen leaves from deciduous trees and shrubs can be gathered at any time of year and stored in a compost bin or cage made from chicken wire. The leaves must be kept damp and the addition of a little manure will provide nitrogen and soil organisms, thus helping to speed up the process along with turning the heap over from time to time. The leaves of evergreen trees and shrubs are unsuitable as they take a long time to decompose sufficiently to produce leaf mould.

Re-potting

Plants will need re-potting probably on an annual basis until they get to a size where growth needs to be curtailed! To maintain healthy growth once this final size is reached, knock the plant out of its pot and carefully remove about an inch of soil from around the root ball using a fork or fingers. Try not to damage the roots too much while doing so. Then simply put a little compost into

Although sempervivums are perfectly hardy outdoors, this lovely collection in clay pots enjoys summer outdoors before being taken back into the alpine house where they can be appreciated in relative comfort over winter.

the pot base and pot up as outlined below.

When using clay pots it is a good idea to place something over the drainage hole to prevent soil from being washed through. This is traditionally a piece of broken clay pot placed upside down, but it could be a thin bit of stone or maybe some broken bits of polystyrene. Plastic pots have drainage holes small enough to prevent compost being washed through. When deciding what size of pot to use, a good rule of thumb is that there should be enough space around the existing root ball to enable light firming of the compost with the fingers. Some compost is first placed in the bottom and the root ball placed on top, firstly having had a few roots teased out if they appear congested. An inch or so of compost is then filled around and lightly firmed, then more compost and firming until it reaches the point on the plant where it did in its previous pot. There should now be around 2.5cm (1in) or so of space before the rim of the pot in which to place a layer of chippings or grit, which will help to conserve moisture. Remember to label the plant and give it a good watering.

Staging

Most growers have staging or benches in their alpine house rather than standing pots directly on the floor, although underneath staging is a useful spot for plants requiring shade or those that are dormant and therefore require little or no light. Staging allows the grower to display their plants at a suitable height for viewing and tending.

As mentioned previously, clay pots are best plunged nearly up to their rims in sharp sand and this of course requires staging that can take the weight. A bed with sand several inches deep over an area of a few square feet will need substantial support. It is always best to play safe and start construction on the premise that 'if it can collapse, then it probably will'! A bench for plastic pots is simpler and cheaper to build as it merely needs to have a slatted top so water can drain through, though it should still be strongly constructed to support the considerable weight of potfuls of heavy, wet compost.

Watering

Knowing when to water or not is a skill that takes time to acquire, though with a few simple tips it should not cause too much difficulty. In any case eventually space in the alpine house becomes tight, and so the occasional plant loss from over or under watering could be viewed as a good thing! If plants are in clay pots plunged in sand, then from autumn until early spring the sand should be kept just moist; the porosity of the pot allows sufficient moisture to keep the soil from becoming completely dry. This means that it is possible to check plants for watering as little as once a fortnight or so, knowing they are unlikely to come to any harm. Remember that some alpines might be dormant in the winter. These are generally summer-flowering bulbous plants, for example *Rhodohypoxis*, and they prefer to be kept dry throughout winter. The first watering in spring will trigger them into growth.

Remember too that the roots of spring-flowering bulbs start growing in autumn and continue through winter and spring, even though growth may not be visible above the compost, so keep them just moist. From then on, when plants are in active growth, there will be a greater demand for water and certainly during hot weather in the summer it may be necessary to water the sand plunge almost every day. This operation is best carried out first thing in the morning or in the evening, thereby allowing a few hours for the plants to take up the water. Plants can of course be watered in the conventional way with a can, and this is often a good idea so that plants requiring lots of water get a good soak, and the occasional wilting plant which has for some reason become dry and requires immediate attention gets rescued! It might be worth mentioning here that a wilting plant does not always indicate a need for water. Rotted roots lose their ability to take up water, thus resulting in flagging leaves. A plant suspected to be in such a condition should have its compost only just moist. This in practice often requires the plant to be repotted. It should also be kept out of direct sun, and somewhere cool. Beneath the staging is often a good place. If the foliage still wilts, try spraying with water. If after a few days there is no sign of improvement then it's probably ready for the compost heap. Remember that some plants, primulas in particular, often wilt naturally in hot weather and will recover once the heat subsides or if given shade.

Alpines grown in plastic pots need a little more care when it comes to watering. As with clay pots, a plant's water requirements are very much reduced from autumn until spring and the best advice during this time as well as during growth is, if in doubt then do not water. Keeping the compost just moist is good practice. Avoid getting water on the foliage if possible as this may lead to fungal diseases such as botrytis, which can cause substantial damage or even death. From spring to early autumn while plants are in active growth, they should ideally be assessed individually for water. All this may seem like a bit of a chore but it soon becomes easy to remember when a certain plant was last watered, and therefore how soon it is likely to need it again.

Maintenance

Besides the usual day-to-day operations of watering and ventilation, it is worth checking plants for general health as a matter of course during each visit. That way any problems can then be attended to as necessary. Mouldy leaves can be a problem especially during dull, damp weather. Carefully pick them off to avoid spreading to other plants, and perhaps place the plant in a better-ventilated spot such as next to an open window or door. It is a common fallacy that dry, warm and sunny conditions are without problems but think again! Red spider mite enjoys such weather and can easily spoil or even kill a plant. Luckily, spraying a fine mist of water discourages the pest and standing plants outside during any rain also helps. Slugs and snails tend to be less of a problem in the alpine house than in the garden due to plants being on benches. Despite being quite an expedition for these slimy pests to negotiate bench supports, if they do succeed, much damage can be done in an evening so take precautions in the form of slug baiting, either natural

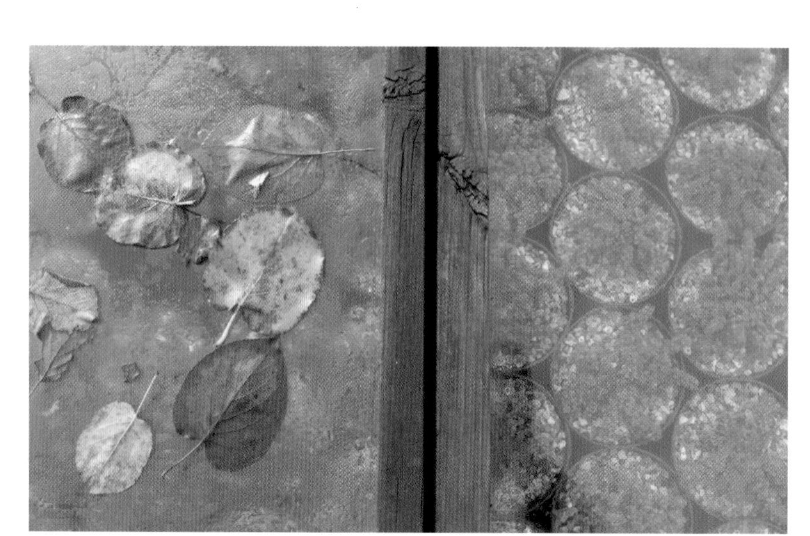

Light levels are significantly reduced by algae and leaves.

or synthetic. Finally, keep an alpine house as clean and tidy as possible. Dirt harbours both pests and diseases and you will be surprised at how much light can be excluded from plants by glass covered in dirt and algae. Algae brushes or wipes off easily when damp, but is almost impossible to remove under dry conditions. These cleaning tasks are best carried out in the autumn, winter or early spring when there is plenty of time to do them thoroughly.

STONE SINKS AND TROUGHS

Many alpine enthusiasts first become acquainted with rock plants by growing them in stone troughs or sinks. These containers are ideally suited for alpine plants as by their nature alpines have relatively little root and modest top growth. This enables a collection of several plants to be grown together in harmony, and with careful selection can give interest all year round. Although in the past stone troughs were not too difficult to acquire, genuine stone sinks are nowadays eagerly sought and as a result both scarce and expensive. Reconstituted stone (i.e. cement or concrete) makes for a perfectly good substitute, being both strong and hard-wearing, not to mention less costly than the real thing. If a challenge is relished, then it is perfectly possible (though messy) to make a cement sink, and this has the advantage of being made to your own specific dimensions. Specialist alpine nurseries sometimes stock them.

A point worth noting is that whether original or replica, the sink or trough must contain a drainage hole and should ideally be not less than about 15cm (6in) deep internally. Less depth leads to rapid drying out during the growing season and little room for roots and compost, though a very shallow sink can be used for drought-tolerant species such as *Sedum obtusatum*. Another option

Home-made containers can be produced to almost any size and shape.

would be to build up a shallow sink with rockwork to enable it to hold more compost.

It doesn't matter whether a sink or trough is in sun or shade as plants can be chosen accordingly. What should be given careful consideration is how the container is supported. Even a modest-sized sink with compost and rockwork can weigh as much as a man, so unless placed at ground level, make sure there is no possibility of it toppling over. Sinks should be raised an inch or so from the surface on which they are placed, using flat tiles or stones made of similar material to the container. If not raised, the drainage hole soon becomes blocked, thereby causing waterlogged compost and sickly plants. Also, if the drainage hole is positioned at one end of a long sink, then tilt the sink slightly that way.

Lewisia cotyledon forms provide long-lasting colour in an old stone trough.

Compost, rockwork and planting

Firstly, place several pieces of broken clay pot or a few bits of thin stone over the drainage hole. This is to help prevent compost from being washed through, or blocking the drainage hole. Next, if the sink is deeper than about 20cm (8in), place chippings in the base so as to leave space for 20cm of compost, which can now go in. This needs to be well drained, and a mixture suitable for growing a wide range of alpines consists of 2 parts loam-based compost such as John Innes No. 2 or 3, 1 part sharp grit, and 1 part leaf mould or substitute such as peat or fine bark. Plants requiring acid compost can be catered for by using roughly equal parts of acid loam, leaf mould and sharp sand. After the first 5cm (2in) firm the compost with the fingers, then add more to within a couple of inches of the rim, again firming.

Unless the intention is to grow one large specimen plant then a few well-chosen rocks can now be suitably placed. Rockwork fulfils a number of valuable functions, in particular improved aesthetics, and the provision of sheltered crevices for plants. It is therefore worth spending some time to find suitable rocks and then arranging them in as naturalistic a style as possible. They could be placed so as to create fissures between, thereby making ideal sites for crevice-dwelling alpines such as *Saxifraga*. Indeed a sink with several different species of *Saxifraga* showing variety of leaf form and colour, even without flowers, can look delightful for the whole twelve months of the year. It's logical to assume that the more rocks used then the fewer spaces there are for plants, but actually in practice rocks, unless they are large, create planting spaces by way of the gaps between them. They also help prevent one plant from growing into another. As a guide to the number of rocks required, a sink measuring 60 × 40cm (24 × 16in) could take as many as 8 to 10 rock pieces (each the size of a fist) along with perhaps 15 of the smaller-growing alpines.

When selecting plants, try to pick alpines that will live happily together and not fight it out with the victor becoming the sole occupant. Cushion plants are often some of the best behaved and always look good together, and there are a great many from which to choose. There are also numerous species of small bulbous plants that can be used to add a splash of colour, such as *Iris reticulata* cultivars, miniature *Narcissus* or the lovely summer-flowering *Rhodohypoxis*. It's a good idea to include one or two plants which will grow over the edge of a trough; perhaps a small creeping *Salix*, or maybe *Silene acaulis* 'Frances' with its dense cushion of bright yellow-green foliage.

Choice of plants

A useful tip when choosing plants for sinks is to avoid vigorous species. Also try to select those that have a long season of interest, whether it is flowers, foliage or both, as the number of plants that can be contained within a sink is relatively small. Otherwise it can be rather a waste of space to plant something which only flowers for two weeks of the year. Possible exceptions to this are small bulbous plants that tend to take up very little space. Of course it isn't essential to plant a varied selection of species together. A single species can be displayed to great effect on its own. Either plant several together to create the appearance of an aged specimen, or a single specimen of a larger growing species can look really effective, for example *Celmisia semi-cordata*. This single-specimen type of planting can be most spectacular, with all the plants flowering simultaneously for a few glorious weeks.

Once alpines have been planted and the compost firmed around, the final touch is to give the whole sink a top dressing of chippings to improve the general appearance. Alternatively, if shade-loving plants have been chosen then a fine grade of bark can often look better. From the plants' point of view it does not matter which. A layer of top dressing not only makes the plants look 'at home', it also reduces evaporation and overheating of the compost, makes pulling out newly germinated weed seedlings easier and prevents compost from splashing onto foliage and flowers in wet weather. A 2–3cm (1–1.5in) layer of top dressing is sufficient, and remember to give the whole sink a good watering once complete.

Long-flowering *Erigeron* 'Canary Bird' in a handmade 'stone' container.

Maintenance

A newly planted sink garden should require only minimal maintenance during the first year or so. Thereafter a few simple tasks are all that are required – the chief being to arbitrate in neighbourly plant disputes! A few snips here and there with the secateurs will prevent one plant from growing into another, although if care has been taken at the selection stage this will be a rare occurrence. Sometimes though, even with the most thoughtful choice of plants, one can prove too vigorous and is best removed to the rock garden or gifted to a neighbour.

If plants are not growing as well as expected or are maybe looking a bit sickly then feeding in some form is probably required. This can be applied in either liquid or granular form. When applying liquid feed, always use half the recommended application rate, and only during the summer months. Granular fertilizer also is applied late spring and summer, following the manufacturer's lower application rate. Remember, however, that if plants are overfed then the result will be an excess of foliage at the expense of flowers.

Although established planted sinks can cope with varying natural rainfall for most of the year, hand watering may be required during the summer months, especially after several days without rain.

The occasional weed seed that finds its way into a sink should be pulled out as soon as it is identifiable as a weed, and well before it has chance to put down a root that may be difficult to extract. Finally, take care also not to allow moss to become established either on the stonework or plants. However pretty it might look at first appearance, it will soon become a menace, so do pull it out.

If you live in an area of very high rainfall, any particular plants that dislike excess winter wet may appreciate some protection. The simplest way is to make a little roof using a pane of glass or perspex attached to some stout wire, thus leaving the sides open to allow good air circulation. The beginning of November is the time to put it in place, and by April it should be safe to remove it. Remember that watering may be necessary whilst the 'roof' is in place.

For plants that may suffer in extremely cold weather, containers can be insulated by being wrapped in layers of sacking or fill sacks with leaves, polystyrene balls or perlite and stack them up and around the container. Even placing some protection on top of the plant or container can make all the difference for delicate plants.

TUFA

Tufa is the name given to a form of porous limestone rock that is often so soft it enables the excavation of holes just large enough for small alpines to be inserted. It is valuable for limited spaces as plants can be grown both in and around it and many of the alpines that grow in tiny crevices on mountain cliffs thrive in the conditions it provides. Holes should be just wide enough to allow the root

Saxifraga 'John Tomlinson' (in bud) growing in tufa.

ball and neck of the plant to be inserted, and can be excavated with a small stone chisel and hammer or masonry bit fitted to a drill. The chosen plant will need most of the compost washed from its roots and is then planted along with a little sandy compost to take up any air spaces. The top can be plugged with a very weak sand and cement mix to conserve moisture. Tufa rock should be either sunk into moist compost or placed in a shallow tray of water to allow moisture to be drawn up by capillary action. Some good plants for growing in tufa include *Draba rigida* var. *imbricata*, *Edraianthus dinaricus*, *Primula marginata*, *Saxifraga* (particularly the small silver-leaved ones and *porophylla* types), and *Silene acaulis* 'Frances'. One problem that may occur is moss growing on the tufa, which eventually spreads among the plants, so for this reason it is important to remove moss as soon as it is noticed.

Artificial tufa

Nowadays natural tufa can be hard to obtain as many of the quarries supplying it no longer have stock or have ceased to extract it. In the absence of natural tufa being available, a substitute can be made using sharp sand (2 parts), medium grade vermiculite (2 parts), peat (2 parts) passed through a 6mm (0.25in) sieve, and cement (1 part). All measurements are by volume. The constituents are mixed together with enough water to achieve a consistency that enables rock-like shapes to be made which, once dry, can be drilled or gouged out to make spaces for alpine planting.

ALPINE PAVEMENTS

A pavement containing alpines could be, for example, a paved area containing sink gardens or raised beds, or a path leading from garden gate to front door. Unfortunately existing paving is only likely to be suitable if it is of the 'crazy paving' type with gaps between. The other problem with existing paving is that it will almost certainly have inadequate drainage beneath it and, whereas it might be tempting to plant among it anyway, success is likely to be limited. A far better approach is to do the job properly and lift the paving, digging out the compacted earth beneath and replacing it with rubble or gravel to ensure adequate

Pritzelago alpina makes lovely dense clumps amongst paving stones.

drainage, then finishing off with a mixture of soil and chippings. The paving can be replaced to leave suitable spaces for plants, which in most cases need only be wide enough to accept the root ball as mainly alpines with a spreading habit are happy covering the stones. Occasional paving stones can be left out altogether if a larger space is wanted.

Making an alpine pavement from scratch involves less work but requires the same good drainage. It could be either of formal design using square or rectangular slabs, or of a more random appearance using stone of irregular shape. Stones or slabs, if laid directly onto soil, should be big enough so that when laid they do not move about or 'rock' when trodden on. If this is likely to be a problem then they should be laid onto five blobs of concrete or cement. Once the paving slabs have been laid then planting can commence. Choose plants with either a creeping or spreading habit like *Thymus*, *Phlox*, *Campanula*, *Helianthemum* or *Sedum*, or those that self-sow between slabs to form close colonies, such as *Erinus alpinus*, *Pritzelago* or *Linaria alpina*. Further suggestions can be found in the plant A–Z listings (*see* page 71).

ALPINE WALLS

Growing alpines in suitable walls gives a lovely opportunity to observe the plants at closer quarters, particularly those with a trailing habit or pendant flowers. Soil conditions in a wall will be extremely well drained which is just what is needed. The aspect of the wall is not really important as long as plants are carefully chosen to suit the particular site – whether it offers sunshine or shade. An essential requirement is that the plants have access to both moisture and food, so the wall must either be filled with a certain amount of moisture-retentive rubble or soil, or be of a retaining wall type of construction with earth behind it. Without adequate access to moisture and soil the choice of suitable plants is limited to a handful of drought-tolerant alpines, namely *Sedum* and *Sempervivum*; whereas with it, the range is enormous. Some of the best types of plants for walls are those with a trailing or mound-forming habit such as *Aubrieta* or *Phlox* as they have the potential to make the most impact. Species with creeping stems are also good, providing the wall does not have mortar between the joints to prevent the plants spreading. Avoid woody subjects as their expanding stems may eventually dislodge stones.

When establishing alpines in an existing wall it is likely that many of the available spaces will be too small to allow easy insertion of the root ball. Thankfully there is a simple – if messy – way of overcoming this, involving a bucket of water and some cold hands. Once a chosen plant has been knocked out of its pot then the root ball can usually be made to fit the chosen space by working it with the fingers whilst in the bucket of water, carefully removing compost rather than roots. The rather soggy result can then be squeezed into a shape for insertion into the wall. Planting is more straightforward when a wall can be planted during construction, though the same method can still be employed. Use a well-drained compost containing plenty of sharp sand or chippings for infilling during wall construction. In all planting circumstances ensure plants are soaked beforehand and that once planted, roots are in contact with soil; they will not grow across a gap of fresh air in order to reach it! Seeds can be sown direct into crevices where there may not be a sufficient gap to accept the root ball of a growing plant. This in fact is the natural method for many plants to colonize walls, for example the fairy foxglove, *Erinus alpinus*. As with any new planting it is important to keep them moist until established, which can be tricky on walls. Once planted, some form of plugging can be used to reduce drying such as sticky soil or a weak mixture of sand and cement. Using a garden sprayer to administer water to the roots, taking care not to wash any soil away in the process, can revive any wilting plants.

Aubrieta 'Swan Red' is a good choice for growing on a wall.

Some particularly good alpines, besides those already mentioned, for a mainly sunny aspect are: *Erigeron karvinskianus, Lewisia cotyledon, Linaria alpina, Origanum amanum, Penstemon rupicola, Saxifraga longifolia, Saxifraga* 'Polar Drift', *Sedum* 'Dragon's Blood', *Sedum* 'Lidakense', *Sempervivum and Veronica prostrata*. For walls mainly in shade: *Campanula betulifolia, Cyananthus microphyllus, Geranium dalmaticum* 'Bridal Bouquet', *Haberlea rhodopensis* (this prefers full shade), *Primula marginata, Ramonda myconi* (prefers full shade), and *Silene acaulis*. This is by no means a comprehensive list and it is worth experimenting as conditions vary enormously from garden to garden and obviously by country also. It is also worth noting that many alpines listed as preferring full sun will actually grow perfectly well with some shade but tend to produce fewer flowers.

If a freestanding wall contains soil and does not have the top finished off with a cap of stone then it also can be planted. The drier conditions encountered here are ideal for alpines such as *Armeria juniperifolia, Delosperma, Crepis incana, Erodium, Origanum, Sedum, Sempervivum, Thymus* and *Zauschneria californica*. For more ideas for both locations refer to Chapter 7.

3 Contemporary settings

MODERN CONTAINERS

Though many people still prefer their alpines to be grown in situations incorporating traditional materials such as stone, others like to develop a more contemporary feel to the garden. There are good reasons for this, including firstly that stone is expensive and often difficult to obtain in some parts of the country, particularly where quarries are few and far between. Stone is also heavy to work with especially when using large pieces. More modern materials such as plastics and metals are available in all shapes and sizes that are more easily transported to and accommodated in gardens. Weight can be an essential consideration for some locations such as balconies and roof gardens. Lastly, though alpines are naturally associated with rock gardens and stone sinks they grow equally as well in containers constructed of alternative materials and their use can offer exciting possibilities, for example the use of silver-leaved plants in a contrasting purple container.

Erigeron karvinskianus is often seen growing between stone steps but here is used to beautiful effect in a metal planter.

Materials

Metal

Most metal planters available are made of a thin gauge material resulting in a light weight, which is fine for small to medium-sized containers but tends to bow out in larger ones unless made of heavier gauge. Finish can be either plain or textured in some way, often raised or indented, and often galvanized to reduce corrosion. The quality of galvanizing varies, with that on relatively cheap objects like plant containers eventually giving way to rust after a number of years. This potential problem can be remedied by painting

OPPOSITE PAGE:
Armeria juniperifolia and *Saxifraga* 'Whitehill' planted in a geometric pattern.

with an appropriate substance. Metals such as stainless steel will not rust and are therefore maintenance free, and make striking objects in themselves but they are often prohibitively expensive.

Remember that any container will need drainage and as not all have holes already drilled it may be necessary to make some. This is easily done with a metal cutting bit fitted to any hand or electric drill – just remember to wear safety goggles and gloves for protection. Compost and planting is carried out in the same way as for stone containers, with a good layer of drainage material at the base being desirable. One possible downside of metal containers is their ability to conduct both heat and cold. Any roots pressing against metal that faces the sun in summer may be liable to damage, and remember that small containers naturally heat up more quickly than larger ones. For this reason it is worth giving metal containers containing delicate plants some shade during the heat of a summer's day. Conductivity of cold is only likely to be a problem if plants of borderline hardiness are grown, in which case insulation could be used as previously outlined in the maintenance section.

Holes have been made in the sides of this painted wooden container to accommodate *Saxifraga cochlearis*. An *Erodium* occupies the top spot.

Resin

Resin is used to make containers that can perfectly replicate stone, lead, wood, etc. in both texture and colour, to give a lightweight and strong finish at a price normally much more affordable than the real thing. Some resin containers we have used at the nursery have shown signs of deterioration, both in terms of loss of colour and wearing of the surface. This is probably due to the continuous freezing and thawing during winter so keeping them under cover at this time might alleviate the problem.

Timber

Wood is relatively cheap and not too difficult to make into suitable alpine planters of varying size. Decking is a very convenient product, with the advantage of being pre-treated with a preservative to prolong its life outdoors, and needs simply to be cut to length and nailed or screwed together. Sheet products such as plywood can also be used though it should be first checked whether they are suitable for outdoor use; if this is not the case then several coats of weatherproofing paint are necessary. Once constructed the container can be left in its natural state or painted in any colour, using paint suitable for outdoor locations.

Plastic

Plastic containers are relatively inexpensive to purchase, durable and lightweight. They come in all manner of shapes and sizes from the humble flowerpot to window boxes, Versailles planters and Grecian urns so there should be something to suit every taste. They do eventually become brittle, but not for several years.

Planting ideas

Herb planter

Usually an earthenware container with planting pockets in the sides, herb planters are widely used for growing a selection of herbs and often positioned outside the kitchen door. However they are also perfectly adapted to growing alpine plants, particularly those that grow naturally on their sides and/or enjoy the sharpest of drainage. A herb planter filled exclusively with plants such as *Lewisia cotyledon*, *Saxifraga longifolia* or different forms of *Sempervivum* makes a long-lasting and spectacular display. Being circular, the herb planter will have aspects that allow both shade- and sun-loving plants to be grown together.

Sempervivums are an excellent choice for herb planters where they appreciate the extra-sharp drainage.

Window box

A window box offers an opportunity for alpines to be grown even if there is no garden space, and offers a choice of aspect: south facing for sun lovers such as *Dianthus*, *Sempervivum*, *Lewisia* and *Sedum*; or north facing for those preferring shade like *Ramonda*, *Hepatica* and certain types of *Primula.* They provide close-up viewing from the comfort of the home – invaluable on cold wet days – and are generally at a convenient height to be easily tended to from outside.

Hanging basket

Traditionally the preserve of annual bedding plants, a hanging basket can also be used for perennial alpine plants, sited in sun or shade depending on the choice of plants. A long season of interest can be maintained by choosing mainly foliage plants such as *Sedum* or *Saxifraga*; otherwise select those with a long flowering season like *Lewisia cotyledon* or *Geranium* 'Ballerina'. Remember to keep hanging baskets well fed and watered throughout the summer, as there is often not much in the way of volume of compost to retain moisture. If new to planting a hanging basket, then it is worth noting that it is usual to make holes in the sides to plant through, and plant more densely than normal so as to quickly achieve a finished result.

Alpines provide an alternative to bedding plants for hanging baskets. *Primula vulgaris* is long flowering and good in semi-shade.

Weird and wonderful
Remember that a planter can be anything that holds enough compost for the chosen plants and has drainage holes or other means of allowing excess water to escape. Sections of old tree fern trunks make really quirky 'pots', along with old-fashioned wooden wheelbarrows, galvanized metal watering cans, old walking boots, shower trays, wooden chairs (with something to hold the compost in place of the upholstery), car tyres... the list is endless!

Sections of dead tree fern trunks make novel containers for drought-tolerant alpines.

Part of the Centenary Celebration display by Stoke-on-Trent City Council at RHS Tatton Flower Show uses *Sedum* and *Sempervivum* planted into a chair seat.

Geometric planting
When planting up a container the plants can be used as objects to create geometric or random patterns. This is most effective when using alpines with a compact and regular shape giving an appearance rather like miniature topiary, particularly evident when highlighted by winter frost. Each individual plant's shape can be further enhanced by isolating it with pieces of slate on edge and this also helps to prevent plants straying into one another's space.

Themed planting
Another imaginative suggestion for planting a bed or container is to develop a theme. As previously mentioned, grouping cushion plants together can be very effective with repetition of shape producing pleasing results. A single colour theme will create mood and atmosphere and could be used to tie in with an already established planting scheme in a garden. Silver- and grey-leaved alpines look very effective when planted together especially if some contrast can be found by choosing those with differing foliage. Alternatively, try growing alpines from the same region or country alongside one another as might be encountered in the wild, perhaps recreating combinations seen when on holiday in the mountains.

Armeria juniperifolia planted in a geometric pattern and neatly divided using pieces of slate.

A silver-themed container incorporating *Celmisia semicordata, Celmisia gracilenta, Celmisia ramulosa* var. *tuberculata, Leucogenes grandiceps* and *Ewartia planchonii.*

RAISED BEDS

A raised bed provides the most versatile site for growing alpines. It can be built to fit precisely the available space and to a height that allows easy access for both viewing and maintenance. Being raised above its surroundings usually means the compost it contains will have improved drainage but only if the soil beneath it is well drained. If not, then it might be necessary to dig out a proportion of soil and replace with hardcore, or perhaps install a simple drain to take excess water away. A raised bed made from timber is relatively quick and simple to build as well as less expensive than brick or stone. Try if possible to use

Phlox cultivars, *Campanula* and *Aethionema* in full flower on a raised bed.

wood that has undergone a treatment process to prolong its life, as being in permanent contact with damp soil soon leads to rotting. The type of decking used for constructing garden seating areas is an excellent timber product to use and comes pre-treated. Timber also has the advantage of being far less likely to attract moss growing on its surface, unlike stone. Stone or brick require a little more expertise as building materials but the process is most therapeutic and should result in an attractive and permanent feature in which to display alpine plants.

Raised beds can have straight or curved sides and are normally between 20cm (8in) and 1m (40in) in height (any higher and there may be problems with stability). Length can be to suit, but careful consideration should be paid to the width so that the centre of the bed remains accessible. The centre of beds up to 1m wide can be reached from either side for maintenance. With anything wider, access provision should be made using rocks or flat stones on which to lean or step on. However it is best not to step on the bed at all because weight on the compost will produce outward pressure on the sides, causing them to bow out or eventually collapse.

If the raised bed is being built using stone, try and choose flattish stones to make the process easier. Resist any temptation to put more than a thin layer of compost between each course of stone for planting spaces, as in time rain will wash it out, making the whole construction unstable. Instead, the main planting pockets can be provided between adjacent stones in a row. The bed is filled as part of the construction process, rather than once the bed walls are complete. This way, a layer of drainage material (rubble) can be incorporated at the base and then built up with layers of compost, ensuring that each stone has either rubble or compost worked in behind to make it steady.

Raised beds built of brick or blockwork are cemented together and can be filled after construction. Spaces may be

Cross section of a raised bed showing how stones are laid to produce an inward lean.

left here and there between the bricks or blocks for planting. It is vitally important that there is good drainage at the base of the bed because if not then the compost will become saturated, causing the walls to bow outwards and collapse.

Making your own garden compost

Grass clippings, prunings, leaves, dead plants, kitchen peelings, teabags and small annual weeds that are not yet seeding can all be turned into lovely garden compost. Cardboard is also a good ingredient but tear it into strips first. There are many compost bins on the market that are a good size for a small garden, or it is possible to make a larger one using timber. Try to make the bin at least 1 × 1 × 1m (4 × 4 × 4ft) as anything smaller won't heat up enough to work properly. More material will generate more heat, which is important for speeding the process of breaking down organic matter. Grass clippings can be added but only in thin layers as thicker ones can inhibit air circulation. The addition of manure, which naturally contains nitrogen and soil organisms, will greatly speed the decomposition process. Digging the heap a couple of times to turn the contents will also speed things up. Compost is ready to use when all organic matter has broken down to produce a dark, sweet-smelling crumbly material. This can happen in as little as 3 or 4 months but often takes longer, depending on temperature.

The compost for a raised bed can be made specifically to suit the type of plants being grown, though a mixture similar to that recommended for rock gardens (as outlined earlier in this chapter) would be fine. Whether the bed is filled during the construction process, or at the end, take time to firm the compost after each 15cm (6in) layer, otherwise the level will invariably sink over time and more compost will need to be added. Finally apply a top dressing; for more information see section on top dressing in the chapter on planting and maintenance (page 40).

Crevice planting containing *Saxifraga cochlearis*.

CREVICE GARDENING

This style is so called because of the way thin pieces of rock are set close together on edge, with only a narrow gap or crevice between them, which is filled with gritty compost. Crevice gardening is particularly suitable for the cultivation of smaller cushion-forming alpines as it helps to provide sharp drainage as well as helping prevent one plant from growing into another's space. If the depth of stone used to make crevices is sufficient, say 15cm (6in) or more, then one plant's roots are kept separate from another so that more vigorous plants find it hard to plunder a smaller neighbour's compost. It would also appear that many alpines, especially cushion types, grow better when their stems and foliage are in contact with rock, perhaps because this keeps them drier by preventing the underside of the cushion as well as the neck of the plant from becoming too wet, or maybe that the drainage is sharper there.

A small-scale crevice garden can be made in a container, whether in an original stone sink or perhaps a home-made wooden box. It might contain only half a dozen pieces of stone but would be ideal for displaying a collection of small silver *Saxifraga* for instance. Alternatively, crevice work could be incorporated as part of a larger rock garden. As for the construction itself, flattish stones of irregular outline often look more natural than ones with straight edges, though neither provides better growing conditions than the other. These flattish stones can sometimes be found at quarries as rejects from crazy paving. Thin stone or slate roof tiles can also work well.

Whereas the compost normally finishes just below the tops of the stones before chippings are added, it could be left a few centimetres lower to provide a slightly shaded area between the stones for those alpines preferring it. In fact, with careful positioning of stones in relation to axis, a crevice garden could have both sun-baked, and cooler, shaded sides. Most alpines suited to crevice gardening enjoy sharply drained conditions and this should be reflected in the compost, with sharp sand and chippings constituting approximately half of the mixture and the rest being a good weed-free loam.

A newly-planted alpine green roof.

GREEN ROOFS

Green roofs have many benefits for both the owner and the environment, beyond their obvious aesthetics. They provide habitat for wildlife, whether simply for birds stopping off to feed on seeds, insects collecting pollen or feeding on honey produced by the flowers, or invertebrates like spiders and beetles that live on it permanently. Larger roofs help to alleviate the effects of localized flooding by soaking up rainwater, thereby slowing its journey to rivers.

Any sort of garden on a roof has challenges not associated with more conventional locations. If it is a roof on a building that can be entered, then it must, of course, be capable of bearing the weight of all materials involved in the construction as well as the maximum weight of water the growing medium can hold. The inherent waterproofing of the roof itself must also be retained for obvious reasons. Therefore the development of a green roof on a dwelling, in particular, requires careful thought and specialist expertise should be sought if such a project is to be undertaken. However, simple green roofs on garden features such as bird tables, rabbit hutches and small sheds are quite within the scope of most DIY enthusiasts. Such an existing roof can usually be adapted without too much difficulty by attaching a wooden edging of between 5cm (2in) and 15cm (6in) depth to accommodate the compost. Next, an extra waterproofing layer is added which also helps prevent any possibly invasive roots from damaging the existing waterproofing.

For a growing medium that is going to accommodate plants, a mixture including materials such as sharp sand, crushed brick, perlite, vermiculite and composted bark can be used to provide sharply drained conditions. If weight allows then a top dressing of chippings or some equivalent will greatly assist in conserving moisture. Plants that are chosen should be able to withstand the often harsh conditions imposed on a roof garden, namely drought and wind. *Sedum* and *Sempervivum* are widely used but there are many more, such as *Phlox*, *Dianthus*, *Thymus*, *Armeria* and *Allium*. Creeping types obviously help to cover the surface and those with a tendency to seed themselves around help to give a naturalistic appearance – alliums such as *A. kansuense* (*sikkimense*) are particularly good at this.

ALPINE LAWNS

An alpine lawn can be made from an existing lawn or area of grass with relatively little effort, though with what some may perceive as some sacrifice to the appearance of the lawn itself. This is mainly due to the required absence of mowing for the months whilst plants are growing. Bulbous subjects are used almost exclusively in these lawns and, with careful planning, can be in flower from early spring until early summer after which it becomes increasingly more difficult to maintain a good show. The season begins with, amongst others, *Galanthus*, *Chionodoxa* and *Crocus* followed by the likes of *Fritillaria meleagris*, *Narcissus*, *Erythronium* and *Tulipa*. The grass cannot be cut whilst the bulbs are in growth, so midsummer is the earliest that mowing can normally be carried out, before any autumn flowering subjects begin to appear.

To establish an alpine lawn it is important that the ground is free from broad-leaved weeds such as *Rumex* (dock), which would, over the course of the few months without mowing, flower and seed everywhere. Planting is easiest in stone-free ground where a bulb planter or trowel can be used, using the opportunity to mix sharp sand or grit into the soil at the base of the hole before placing the bulb on top and replacing the plug of soil and turf, ensuring that the bulb is at the correct depth. When planting in stony ground it is easier to use a spade to lift a section of turf approximately the same size as the spade's blade. Then dig over the soil, removing any larger stones and adding sharp sand or grit before planting the bulbs and replacing the turf. Hopefully under suitable conditions a number of species will self-sow to give a naturalistic effect but remember that most bulbs will take a few years to reach flowering size from seed.

Chionodoxa luciliae can be naturalized in an alpine lawn.

4 Planting and maintenance

PLANTING

Generally speaking there are usually six months of the year when conditions are suitable for planting in the garden – whether it be in a rock garden, scree garden or containers. Planting can commence as soon as the weather improves in spring, usually around mid-April in an average year, when hopefully we have seen the last of all but the mildest of frosts and the ground is beginning to warm up a little, thus encouraging alpines into growth. Planting may continue through the summer and well into autumn, but really the end of September should be the finishing point as by October the ground is cooling significantly, and alpines planted at this time or later may not establish before the onset of winter.

Spend time planning

While it is tempting to just get started without giving much thought to an overall plan, it pays to spend some time trying to picture what the garden will look like in a few years' time. If using dwarf conifers or other shrubs, check carefully on their likely dimensions after say five years, as many so-called dwarf conifers get far too big for all but the largest rock gardens. They can look out of scale with alpine plants, and their rate of growth means that they quickly swamp the smaller species or cast shade, sometimes on plants that do not like it.

OPPOSITE PAGE:
Planting a sink garden with saxifrages.

Chamaecyparis lawsoniana 'Green Globe' – a four-year-old plant in a 13cm (5in) pot next to a mature fifteen-year-old specimen.

How many to plant?

You may wish to plant for overall effect and have some plants in small groups thereby producing a naturalistic drift. This is a good idea with very small plants, such as *Saxifraga cochlearis*, whose little cushions of foliage look very effective planted along a crevice for example.

Larger, more spreading species are generally planted singly as, by their nature, they provide more impact. Plants with a creeping habit also need only be planted singly, though of course more impact would be gained by planting say three. However, watching your alpines grow is all part of the fun, and not allowing for future development means that you might well have to cut back or dig out plants sooner than anticipated. As a rough guide on how many to plant in a given space, incorporating small as well as larger plants and assuming rockwork has been used, then five to seven per square metre would be a good starting point. Allowing plenty of space for growth also gives spare planting ground for the inevitable future purchases which we all find irresistible from time to time! And remember that if a plant fails to

The roots of this dwarf conifer can be carefully teased out using a fork to remove it from the pot.

grow in the way you anticipated or simply just doesn't look right then most alpines can be happily transplanted and tried somewhere else.

Sun or shade?

It is very tempting to choose plants that you simply like the look of initially, but it is important to know what conditions the plant favours as this will determine whether or not it is suitable for your situation. If your site has both sunny and shady positions then remember to choose suitable plants for those conditions.

Planting technique

Firstly make sure the plant is well watered, as a dry plant will already be under stress and being knocked out of its pot and planted in a new environment will only add to its discomfort. Assuming the plant is pot grown, first carefully turn it out of its pot by pressing on the base of the pot. Remember to keep one hand over the plant itself to prevent it falling on the floor. If gentle pressure is not enough to free the plant from the pot then tap the edge of the pot on a solid surface, while holding the pot and plant upside down. Look carefully at the root ball, and if the roots are so dense as to make any compost invisible, then carefully loosen the root ball with your fingers or, if they are very matted, then gently tease out some roots using a kitchen fork or similar tool. This process helps greatly in encouraging new roots to grow out from the previous confines of a pot. If on checking the root ball you can see a fair amount of compost, then this process can be omitted.

Planting now simply involves digging a hole large enough to accommodate the root ball, and once the plant is in the hole, filling around it with soil/compost. Finish off by firming it in with your hands, making sure that the surface of the original pot is at the same level as the garden soil, so as to leave space for a layer of top dressing between plant and soil level. The depth of top dressing should be a minimum of about 2cm (0.75in) but 3–4cm (1.5–1.75in) would be better still. Don't worry if you seem to have buried your plants a little, they will soon grow up through the top dressing, but try not to leave them sticking up, as this exposes the compost surface and can lead to loss of moisture. The plants should now ideally be given a good watering, which helps to settle the compost around the root ball, and of course provides moisture, thereby minimizing any extra stress to the plant.

A tight fit

It often happens that the plant you have chosen has too large a root ball to be planted in the allotted space, such as between wedges of upright rocks or particularly when planting into walls; fortunately this problem is easily overcome. Empty the plant from its pot and work the root ball gently with your fingers in a bucket of water to remove as much of the compost as is necessary to allow it to fit the crevice or crack. The root ball can be squeezed and flattened into a shape to make it a perfect fit.

Top dressing

The final appearance of an alpine bed can be greatly enhanced by the addition of a suitable top dressing, generally in the form of some sort of gravel or chippings. A top dressing has several important functions beyond purely visual: it helps prevent the surface of the compost forming a crust which can disrupt water penetration; it greatly reduces the loss of moisture – particularly important both on sunny and cold windy days; it provides insulation from the heat of the sun as well as from

Many different sizes of top dressing material are available. The lower right hand pile shows the other three mixed together, which often produces a more natural finish.

cold; weeds that germinate are more easily pulled out; splashing from heavy rainfall is prevented thereby keeping flowers and foliage clean; and the 'necks' of plants are prevented from resting on wet soil, so reducing the chances of rotting.

The choice of top dressing will to some extent be dictated by availability. Garden centres often stock a range of grits, though some of their offerings may be more suited to the churchyard. If you have used stone in the construction of your alpine bed then try if possible to use a material that blends in with this stone. The size of grit or gravel is not so important, but the application of two or more different types and sizes tends to give a much more natural appearance. A thin covering is better than nothing at all, but do try to provide a layer 2–3cm (0.75–1.5in) deep. Application of top dressing can be carried out either before or after planting. Simply scrape the top dressing to one side before digging the hole if you are planting after a top dressing has been applied. Make sure that whatever type you use it is spread under any foliage, right up to the main stem.

Labelling

Labelling is a matter of preference, but ideally you want a marker for the plant's position (especially if it is one that goes dormant over winter) and for its name to be clearly legible. A simple plastic label poked a couple of centimetres (1in) into the soil will probably, over the course of a few months, have been either pulled out by birds, lifted by the frost and blown away, or become illegible due to the effects of sun and rain, as the majority of so-called 'permanent' marker pens are anything but. A humble pencil mark, however, remains completely legible on a plastic label. If the label is inserted with just enough above ground to remain visible, then it is far more likely to remain in place. If you feel that labels looks unsightly, then try burying them at a distance and orientation from each plant, which can be applied consistently to all your plants, so you know where to dig to find them! For example, put all labels 10cm (4in) away from the stem and in a southerly direction.

AFTERCARE AND MAINTENANCE

Your alpine plants will require only a little aftercare during their first few years.

Watering

Ample watering is essential to ensure successful early establishment, as new plants

Cutting back straggly growth on *Aubrieta* produces a more compact plant.

initially tend to have a limited root system, and are therefore unable to access water from more than a few inches below the surface. For the first few weeks after planting, check beneath the mulch as to whether watering is required, particularly during periods of sunny and windy weather, which are the conditions most likely to dry the soil. This is especially important if planting has been carried out during the summer. Planting in spring or early autumn, when conditions are less extreme and there is likely to be more moisture, may help if you live in an area prone to periods of drought. When watering, remember to try and do it when evaporation is least likely to occur, such as early morning or evening.

Cutting back

You will doubtless find that some plants are more vigorous than expected, and that they begin to encroach on others. This is quite normal. Simply dig up either the encroacher (or the encroached upon) and move it to a new position. This is often a better approach than trimming away at shoots as this tends to encourage further growth. That said, several alpines respond well to cutting back, either on a regular basis or every few years. *Helianthemums* are a good example, but try to avoid giving them a severe haircut every year and removing all growth except for the last few centimetres. A better approach is to try removing around a third to half of the growth annually, in spring, and then cutting back the untrimmed growth the following year. This tends to maintain the plant at a manageable size, whilst still encouraging plenty of new flowering growth, and minimizes untidy straggly stems. Plants such as *Aubrieta* are best cut hard back once they become straggly. Late summer or early autumn, once flowering has finished, is a good time to do this as the plant then has enough time to make new growth that will begin flowering again in the spring.

Feeding

Growth in the first few years after planting should be such that no additional feeding will be required. However, if your plants begin to look unhealthy, say showing yellowing foliage, or are failing to make much growth, then the application of a slow-release fertilizer over the whole of the area could be carried out. Use a readily available product such as blood, fish and bone, or bone meal, at the lower end of the rates recommended on the packet. Spring or summer is the best time to carry out this feeding. Avoid doing it in autumn, which may encourage a wealth of relatively tender shoots that may then be damaged by the frosts. Once the feed has been sprinkled over the soil surface, work it into the top few centimetres. Such a treatment should last for the whole season. Whilst wearing protective gloves wash off any residue from foliage as this is can cause burning of the leaves, which can lead to further problems.

As an alternative, if you have access to good weed-free garden compost or leaf mould, then use a 2.5–4cm (1–2in) layer of this, first mixed 50:50 with grit or sharp sand, and work it around the plant, after first scraping the top dressing aside.

Liquid feed is a convenient method of feeding, and has the advantage of almost instant results. Certainly plants can be 'greened up' within a couple of days of application, although obviously any growth will take longer! Most popular brands of fertilizer work well, although the majority are aimed at hungry perennials or vegetables so it is best to adjust the dilution rate to roughly half that recommended. Simply apply it directly over the foliage using a watering can fitted with a rose, remembering to avoid hot, sunny weather as this can lead to scorch. Application in the early morning or evenings is therefore best. Late spring and summer are the times to apply liquid feed, and as the application is quite weak, it can be performed several times during the growing season.

A simple application of top dressing soon refreshes an alpine display.

Replanting and replenishing soil

Though some plants live for tens of years some losses are inevitable, whether as a result of natural lifespan, pest or disease, being swamped by neighbours or, as is sometimes the case, they just seem to 'give up'. Losing plants has a positive side though as it creates a space for one of those plants every gardener has 'waiting to be planted somewhere'.

If the planned replacement is of the same species, just make sure that the original loss was as a result of something other than a pest or disease specific to that plant, because these can remain in the vicinity ready to strike again. Whatever the replacement, take the opportunity to replenish the compost unless it is less than a year or two old. This means digging out the compost to a depth and width containing any of the old roots, forking over the base of the area to maintain drainage and then adding new compost for the new plant. If this process is carried out on an 'as and when' basis then eventually almost the whole area will have fresh compost over a period of several years. This is important because not only does the ground eventually lose fertility naturally, but the action of being walked upon for maintenance, along with the forces of gravity and weather will, over the course of only a few years, begin to lead to compaction of the soil with the resultant loss of drainage and aeration.

Containers and troughs, if planted correctly, will last for a number of years, depending on which alpines are used, before they begin to look as though they need an overhaul. This is due to nutrients being exhausted and compaction from weathering. To some extent this can be delayed if fresh compost is introduced whenever replacements are required, as practised in the garden. However it is often simpler in practice to dig out any plants that are going to be retained, remove the old compost and replace it with new, before replanting the container and adding top dressing.

Replenishing top dressing

What may come as a surprise is the gradual disappearance of the layer of top dressing. The explanation is simple – it results from the constant actions of earthworms, the forces of gravity and general planting or weeding actions. It is therefore necessary to top up the top dressing from time to time. It is surprising what a difference a mere sprinkling of clean new gravel can have on the appearance of a rock garden or planted container. Aside from the instant visual makeover, new top dressing also helps to discourage moss (the bane of an alpine gardener's life) which can, if left unchecked, severely damage if not kill some of the smaller cushion plants. So always try and remove moss as soon as you notice it, however attractive it may sometimes appear!

Weeding

A few weeds are always likely to appear, either from seeds blown on the wind or indeed those that have been brought in with compost. Hand weeding, if done regularly before weeds have a chance to become established, should not take too much time and can be surprisingly therapeutic. Try if at all possible to remove weeds before they can seed, otherwise the task can become even bigger. Annual weeds should not be too much of a problem to remove by hand but perennials such as nettle, dock and dandelion seem to have evolved to frustrate the alpine gardener by choosing to grow right through some tiny treasured cushion plant or indeed next to a favourite specimen that is wedged between two immovable rocks. In this instance, short of dismantling the rockwork, the offending weed could be dabbed with a translocated weedkiller but do take great care not to get any on surrounding plants. A plastic collar around the treated weed helps to avoid contamination until the weedkiller has dried.

Weeds

It is important to control weeds in the garden as they compete against cultivated plants for space, food and water and may harbour pests and diseases. New areas to be planted should be clear of weeds as it is always more difficult to remove them afterwards. The key with weeds is to identify them in their early stages and understand their mode of growth. A golden rule is to prevent all weeds from producing seeds because they can then be quickly and widely dispersed. Annual types such as hairy bittercress (*Cardamine hirsuta*), annual meadow grass (*Poa annua*) and chickweed (*Stellaria media*) live for only one year, and therefore they are easily eradicated if prevented from seeding by hand weeding, hoeing or spraying. Perennial weeds like dock (*Rumex sp.*), creeping thistle (*Cirsium arvense*) and perennial nettle (*Urtica dioica*) can be controlled in a small area by simply digging them out. Their tough roots can sometimes make complete removal difficult and may require the use of a translocated herbicide that moves through and kills all parts of the weed including the roots.

If you are growing alpines anywhere other than the driest parts of the country then you will soon become familiar with moss, which if left unchecked can soon completely swamp smaller alpine plants and can be very hard to remove. Control is problematic and should definitely be tackled as soon as possible. Never mind how attractive a few tufts of moss look on the edge of an old sink garden – they will soon propagate by spores in their thousands and leave no cushion plant untouched. You have been warned. Winkling the little tufts out with some sort of pointy tool works best. Moss-killing chemicals are available but should be tested very carefully first as some alpines are highly sensitive to them.

Winter protection

Alpines in their native mountain habitat are naturally exposed to extremes of temperature with the air and ground below freezing for many months. Domestic alpine gardens in most cases also experience freezing weather particularly during winter and spring, but there is one significant difference between a winter in the mountains and one at lower altitudes and it is responsible for the majority of winter losses. In mountain habitats, when snow falls in the autumn, temperatures are such that it is too cold for melting to occur and so each subsequent deposit of snow adds to the last. The alpine plants beneath are dry and protected from the desiccating effects of the icy winds above until spring. This contrasts with the alpines in our gardens at lower altitudes, which for several months have to endure freezing and thawing with little or no snow cover. This can lead to a perpetual state of soggy soil, which is the one thing alpines especially dislike.

A simple square of glass supported on wires can provide protection from excess rain and snow.

However, a great many rock plants are sufficiently robust to endure a wet winter, and a gritty well-drained soil coupled with a mulch of chippings around the necks of plants helps them to achieve this. Those that can suffer in a wet winter are usually given the protection of an alpine house or frame where soil moisture content can be controlled to some extent. However, we all like a challenge and for those plants that are better kept a little drier then it should not be too difficult to provide a roof of some sort over them – perhaps along the lines of a piece of glass or clear plastic supported on stout wires, remembering to leave the sides open for air circulation.

ALPINE CARE CALENDAR

Here is a guide to possible jobs that may be carried out during the alpine gardener's year. Seasons vary as do gardens and so there are no hard rules as to what can or cannot be done in a particular month.

January

- This can often be an extremely cold month, with snow, so keep an eye on snowfall and brush off cloches, polytunnels and alpine houses as the weight can cause such structures to collapse.
- Keep an eye on plant labels outdoors as the frost can lift them from the ground, making subsequent identification difficult.
- Weather permitting, January can be a good month to undertake any hard landscaping/construction work in the garden, preferably while the ground is dry.
- Carry out any pruning and look out for damage to small shrubs from the weight of snow. Prune away any damaged stems.
- Take root cuttings from geraniums.

February

- Continue with many of January's tasks.
- Seed sowing should be completed by now to take advantage of any cold weather. Some seedlings may be large enough to be pricked out now but only do this if you have a greenhouse in which to keep them, offering some protection should the weather become very cold.
- If the weather permits, try a little weeding to make the task easier as the season progresses.
- Birds can often take a liking to brightly coloured early flowerers and peck the petals off. Unfortunately there is very little that can be done to discourage them except to place protective net or wire mesh covers over the plants.
- Propagate plants such as *Antennaria*, *Astilbe*, *Oxalis* and *Thalictrum* by division.

March

- Start visiting your local alpine nurseries and flower shows for spring colour ideas.
- Commence planting, depending on the weather, but bear in mind that it can still turn very cold, so give protection to anything that is newly planted.
- There is still time to sow seeds of alpines that don't require a long cold spell, such as *Campanula*, *Geranium* and *Erigeron*.
- Soft cuttings may be found on a few early growers in the alpine house such as *Campanula*, *Cyananthus* and *Silene*. These are normally removed as short basal growths.
- Alpines in the garden can be given a good start to the season by feeding with bone meal or blood, fish and bone, lightly forking it into the soil surface. Alpines also enjoy a top dressing of gritty compost every couple of years.
- Divide clumps of *Rhodohypoxis* that have become congested, potting them in acid compost and giving them some water.

April

- Remove any remaining glass protection covers once the threat of snow has passed.
- Shading of alpine houses and frames may be necessary if the spring weather is warm, as the sun can be surprisingly hot and easily scorch young plants.
- Trim any lawn edges from alpine beds.
- Remove any old dead growth from the previous year to allow air and light to reach new shoots.
- Keep an eye out for any pests in the greenhouse or cold frames especially aphid outbreaks.
- Continue to make new plantings.
- Reposition any plants that are not in the right place, due to crowding neighbours, clashing colours, etc. This is best done before growth becomes too advanced and causes extra stress from moisture loss due to root damage.
- Many plants will have vigorous new shoots ideal for propagation.

May

- Often alpine gardens are beginning to look their best during May so make a note to visit your local (or not so local) one. A list of alpine gardens and nurseries is given on page 135.
- Apply shading, whether as netting or whitewash, to any cloches, propagation facilities or glasshouses. If netting is used, try to remove it on dull days.
- Aphids can be a nuisance during the spring and summer months so keep an eye out for them and use a preferred method of removal – a soapy solution, eco-friendly spray, or if numbers are few by simply squashing them.
- Collect any ripened seeds, storing them dry in envelopes in a cool place until needed. Some seeds such as *Hepatica* and *Pulsatilla* are better sown fresh, so sow these now. Continue collecting seeds until the autumn.
- May is a very busy month for propagation by cuttings. Also continue to prick out seedlings before they become overcrowded, making them difficult to separate.
- Continue visiting the late spring and early summer flower shows and Alpine Garden Society Shows to buy unusual plants, gather tips and to see the wonderful displays.

June

- Regular watering may be needed from now through to September, especially for plants in containers in full sun. Watering is best carried out in the evening or early morning when it is cooler and the water is less likely to evaporate.
- Cut back any plants that are becoming a bit too large and may be overshadowing other smaller plants around them. Otherwise, you may need to consider moving the smaller plants to give them a better chance of survival.
- Propagation continues.

July

- Continue watering as required. Containers may need watering twice a day if the weather is really warm. Remember that *Rhodohypoxis* will go

dormant if they dry out so keep well watered, even standing in a shallow tray of water during hot periods, to keep them flowering for as long as possible.
- Take cuttings of new shoots on *Leucogenes* and *Erigeron* 'Canary Bird'.
- Deadhead plants such as *Epilobium glabellum* and *Lewisia cotyledon* to prolong flowering.
- Move any lightweight containers around in your garden to give prominence to the best ones each month, and to create different views.
- Collect seeds and either sow them or store in labelled paper packets in a cool, dry place until winter.
- Many spring-flowering bulbs will have finished growing; this is indicated by brown foliage. Now is an ideal time to dig up, divide and replant them. Trim back any plants that have become straggly such as *Aubrieta* and *Helianthemum*. Cut back hard they will reward with increased flowering the following year.

August

- Maturing growth provides heel cuttings and the more woody cuttings.
- Alpine lawns can usually be cut now before any autumn-flowering subjects begin to grow.
- Continue to collect seeds as they become ready.
- Keep the alpine garden tidy by trimming or pruning back plants that have finished flowering, unless they are needed for seed.

September

- September is usually the last month with enough ground warmth for planting – early October can work too, but not in cold areas.
- Take cuttings of rosette-forming *Saxifraga* and semi-ripe cuttings of anything missed in summer.
- Carry out any pre-winter weeding of alpine beds and containers.
- Shading is now usually only required on propagation frames around midday when the sun is at its strongest, so remove the shading at other times.
- Plant alpine bulbs.
- This is a good month to get on with building new raised beds, screes, and rock gardens.

October

- Remove any fallen leaves, as they block valuable light from alpine plants.
- This month is definitely the last chance to get any planting done before the soil really gets too cold!
- Plants benefiting from frost protection should be given either a good mulch of bark or chippings to protect resting dormant crowns, or taken into a cold greenhouse or garage.
- Place covers over any plants susceptible to winter wet.
- Remove algae and fallen leaves from cold frame and greenhouse glass to maximize light to plants beneath.
- Remove any shading from greenhouses or cold frames.
- Order alpine seed catalogues or view online – the Internet is one of the best places to source suppliers.
- Take hardwood and heel cuttings, also make cuttings of *Saxifraga* rosettes.
- Finish planting bulbs.
- Continue building new beds but wait until spring for planting.
- Dismantle any unprotected watering fittings before they are damaged by frosts, remembering that expanding ice can crack metal!

November

- Keep removing fallen leaves from plants.
- Keep an eye on the weather as temperatures can drop very suddenly and seedlings need protection.
- Continue constructing beds, etc.

December

- Remove any leaves and fallen branches after strong winds.
- Plan an alpine holiday. There are specific plant-seeking trips to many alpine regions where a guide who knows the area will be able to take you to, or advise on, the best places for alpines. Alternatively do a little research as many ski regions keep resorts open during summer including lifts, thereby allowing easy access to mountain slopes.

5 Propagation

THE THRILL OF MAKING ANOTHER PLANT

Sooner or later, either by accident or design, you will find it hard to resist the temptation to propagate some of your plants. It is an absorbing and rewarding aspect of growing alpine plants and there are several good reasons to 'have a go'. Favourite plants can be shared with friends (remembering of course that should you happen to lose your original plant, then you can always ask your friend for a cutting). Plants can be quite expensive when purchasing the quantities needed for a large rock garden for example, so a good saving can be made even if just a handful of plants are produced for free. But many people simply enjoy the thrill they get out of making another plant, that excitement often being relative to how difficult a plant is to encourage to make roots and become established. It can become an addiction!

Alpine plants are propagated either sexually, that is by seed, or vegetatively, known as asexually, involving the removal of part of the plant – be it a shoot, leaf or root. It is worth pointing out here that plants produced vegetatively will be identical to the plants from which they have been removed, being in effect, clones. Named varieties must be propagated in this way in order to maintain their characteristics. A plant grown from a seed usually has the appearance of the plant from which it was collected but has the potential for variation – for example flowers of a different colour or more compact growth. These characteristics have long been exploited by gardeners and nurserymen alike and who knows – there is always a chance of a *Pulsatilla* seedling which flowers from spring to autumn!

OPPOSITE PAGE:
Saxifraga longifolia grown from seed.

A seed tray with a plastic lid can make an effective mini propagator.

Seeds can in most cases be germinated out of doors with little or no protection at all though a cold frame can be useful in affording some protection from, for example, lashing rain which can dislodge tiny seedlings or the drying effects of wind and sun.

FACILITIES

Alpines can be propagated with the simplest of facilities and whereas it is always nice to have the latest technology at your disposal (a mist unit or tissue culture facility, for example), it is worth making the point that what often makes the difference between success or failure is not how up-to-date a propagator is but how well it is managed – that and, of course, experience! Heated propagators and misting systems are very nice to have but there are no alpines that cannot be rooted without one. They merely extend the season of propagation and/

or make the cuttings easier to manage. So remember that familiarity and experience with a chosen propagation facility is worth a great deal, and likely to be more important than the type of facility.

Cold frame

For the first five years at our nursery all cuttings were rooted in a cold frame, and with good results. A cold frame mainly offers protection from rain and wind, and provides a humid environment with a temperature above that outside. This provides suitable conditions for the rooting of alpine plant cuttings as well as providing somewhere for newly rooted and potted cuttings and seedlings to establish while they overcome the shock incurred during the potting process. A cold frame can be situated in either sun or shade. A shaded one will be easier to manage as it avoids the high temperatures caused by direct sun, although rooting does take longer than in a sunny frame.

The shaded cold frame is perhaps the best choice for people who are generally away from home for most of the day. Management is minimal – it simply requires observation to determine whether some ventilation is required in order to avoid too humid an atmosphere, which can lead to fungal problems; conversely not enough humidity causes wilting of the cuttings which quickly leads to failure. As a rough rule of thumb, try to remember that increased temperatures require greater humidity.

A cold frame situated in a sunny spot naturally warms up rapidly if the sun is shining or there is thin cloud, and un-rooted cuttings can quickly lose water, wilt, and die if heat is excessive. It is vital that steps are taken to avoid this happening, and some form of shading is necessary, usually in the form of netting, which could be attached to a lightweight wooden frame to make application or removal easier. As already mentioned, an excessively dry atmosphere in the frame can cause stress for the cuttings. This can be remedied by applying moisture to the inside of the cold frame via a watering can fitted with a rose, avoiding over-wetting the plants. The moisture will slowly evaporate, thereby raising humidity and preventing delicate cuttings from wilting.

A useful gadget for the cold frame is a thermometer that records maximum and

Reducing a cutting's stress

The moment a cutting is removed from a plant it has of course no roots, and therefore is vulnerable to drying out until the cut stem is placed in the moist cutting compost. Maintaining it in as near to ideal conditions as possible reduces the stress for it and results in roots developing as soon as possible. The main factors influencing stress are light, heat, rooting medium and humidity. Each has its own optimal level and parameters for problems. Though cuttings like good light levels remember at all costs to avoid excessive build-up of heat in any propagator due to sunlight falling directly on it, except perhaps in winter, although even then temperatures can quickly soar beyond 27°C (81°F), which is around the point above which a cutting with no roots will suffer. Thankfully, heat build-up can be avoided simply by shading a propagator, though always try if possible to remove any shading during dull weather. Remember also that higher temperatures will necessitate an increase in humidity in order to maintain a cutting's turgidity, so it is important to have a propagator without gaps, which would allow moisture to escape. Any excess moisture within the propagator can be got rid of simply by lifting the top up for a while. This will also quickly allow excess heat to escape.

A thermometer is very useful for monitoring temperatures in a cold frame or other propagation facility.

minimum temperatures. It provides a useful guide to help you achieve the ideal temperature of 18–20°C (64–68°F). Anything much above 25°C (77°F) is best avoided. For several months of the year, however, such ideal temperatures are impossible to achieve, and though rooting is still possible, it simply takes longer. Artificially raising the temperature is an option by means of electric soil-warming cables controlled by a thermostat. Even maintaining a winter minimum of around 5°C (41°F) will make quite a difference. Insulating the cold frame is a good idea especially if it is to be used in colder weather. This can be done by using 5cm (2in) thick polystyrene sheet at the base and perhaps sides of the frame, with bubble polythene fastened to the inside.

Cold frames can be purchased ready-made or can be made without too much difficulty. In terms of construction it should have the following: a sloping roof to allow condensation to run off and not drip on to cuttings; freedom from gaps, which would allow loss of humidity; have a top of glass, clear plastic or polythene which can be easily lifted off or fastened back for access, and propped up by varying degrees for ventilation. The frame should be placed on a surface that allows any excess water to quickly drain away, ideally gravel, which to some extent discourages cuttings from rooting through but which retains some moisture, thereby helping to maintain humidity. Despite the simplicity of a cold frame it offers the opportunity to propagate alpine plants with relatively little expense and effort and would be the first choice for an amateur propagator.

Greenhouse

As a greenhouse generally contains areas with differing temperatures and shade values it offers the gardener the chance to have a number of different facilities. Under the benching will be the coolest spot and provided that getting down on hands and knees to tend cuttings is practicable, it makes an ideal spot for a cold frame type construction or simply a seed tray with a plastic lid. Just make sure sun does not shine directly onto it, or provide shade if it does, and here it will be possible to root a wide range of plants, particularly those which like to grow in some shade such as *Ramonda*, *Hepatica*, and many species of *Primula*.

Those plants requiring higher levels of light can be rooted in some sort of frame at bench level or, as with under the bench, a seed tray with a clear plastic lid on it can be used, remembering to shade if necessary. Even the humble polythene bag can be pressed into service as a means of maintaining the humidity a cutting requires.

Mist unit

A greenhouse also offers convenient accommodation for a mist unit, which can be either situated on a waist-level bench or at ground level, with a layer of sharp sand containing electric soil-warming cables and a thermostat set to maintain a temperature of around 18–20°C (64–68°F). Seed trays or pots containing cuttings are placed directly on top, thereby achieving the same optimal temperature. The mist unit itself comprises a sensor that detects any moisture on a surface replicating a leaf. A dry surface triggers a mist of water via nozzles situated 45cm (18in) above the cuttings which falls onto their leaf surfaces, reducing water loss. The moisture is instantly detected by the replica leaf, which then stops misting.

Soil-warming cables

The use of electric soil-warming cables used in conjunction with a thermostat in a bed within a greenhouse or polytunnel enables ideal temperatures to be maintained in the propagation environment (most beneficially in the rooting medium), and they are best used in conjunction with a thermostat. Although temperatures in the range of 16–20°C (61–68°F) are sensible to aim at for rooting soft or semi-ripe cuttings, such temperatures can be expensive to maintain throughout autumn, winter and early spring. During these times the rooting of semi-ripe and hardwood cuttings is easily achievable even with the thermostat set to a minimum of around 5°C (41°F). In fact many alpines

Soil warming cables in a sand bed which can be controlled by a thermostat. The sand has been removed to expose the cables for purposes of illustration.

are most easily rooted this way as cooler temperatures impose less stress on the cuttings. On the other hand, cuttings take longer to root – frequently several weeks or even two or three months – rather than the couple of weeks which can be expected of soft green cuttings taken at the height of summer. The use of soil-warming cables in a bed of sand should be used in conjunction with some form of light-transmitting cover. This is easily achieved by a polythene cloche type structure, which will then help to conserve some humidity and heat. During cold winter weather, electricity consumption can be reduced by the use of bubble polythene – two layers still allows sufficient light and reduces heat loss further still.

A covered bed containing soil-warming cables is perhaps the most versatile type of propagator and has the benefit of being relatively inexpensive and easy to install. For detailed information on how to make a propagator incorporating soil-warming cables and/or a mist unit the Internet is the best source.

Windowsill

A windowsill in the home provides space for propagation although other members of the household may not appreciate this! If family members are in agreement, then a small electric propagator (the sort that has just one seed tray) should not be too obtrusive and will enable the rooting of dozens if not hundreds of cuttings over the course of the year. Should space be restricted then even a pot with a polythene bag over it can be used. Just remember that if you are using anything other than a window with a northerly aspect, overheating can soon occur when the sun shines, so some form of shading would be necessary.

PROPAGATION METHODS

There are methods of propagation to suit all levels of interest, from the simple division of a plant, which needs only a moderate amount of skill, to raising from seed which also requires a medium level of care and attention, through to tissue culture, which pretty much necessitates laboratory conditions and very careful attention to detail and therefore falls outside the scope of this book.

Raising alpines from seed

It is worth considering the pros and cons in the raising of alpines from seed. Firstly it offers the grower a method of raising a large number of plants for relatively little outlay. This is particularly relevant should the plant chosen for propagation have only a few shoots available from which to make cuttings, whereas if the same plant produces seed, then there are often several dozen if not hundreds of seeds, each with the potential to become a plant. It is worth mentioning here that when a batch of one particular species of alpine is grown from seed it is often noticeable that they are not identical – variation being shown in, for example, leaf size or shape, flower colour

Pulsatillas are best raised from seed sown as soon as ripe.

Collecting and storing seed

Carefully collected and stored seed will give the best germination results. Some species retain viability for several years if stored well and so it is worth the time and effort involved to do so. Seeds should, if possible, be collected in dry weather for if the seedhead is damp from rain or dew then this moisture might lead to fungal problems later. There are a few alpine plant species where seed can be harvested still green but for the vast majority it is ready when the capsule containing the seeds becomes brown, dries and splits to reveal what are usually brown or blackish seeds. If they rattle around and fall freely from the seedpod then put them straight into a small paper bag or envelope made specifically for seeds and, most importantly, label the packet with the name of the plant and date collected plus any other relevant information. Sometimes it may not be possible to collect seed under dry conditions, in which case place the cut seedheads on sheets of paper under cover until dry enough for the seed to fall from the seed capsule. It might be necessary to check the condition of seedheads on a number of consecutive days for ripeness as *Geranium* species, for example, have an explosive mechanism for distributing seed, resulting in them being catapulted out of sight more or less as soon as ready. Do not store seeds in a polythene bag, as any moisture either on the seed or in the bag is likely to cause fungal problems. Packets of labelled seed can be kept in a plastic box with a tight fitting lid containing a desiccant and stored somewhere cool such as a garage or kitchen fridge.

or perhaps a combination of these. Such deviations from the norm are worth looking out for, as this is one way in which new varieties can occur. Growing a large quantity of any one particular species from seed will generally, sooner or later, result in some particularly outstanding feature, which makes it worth maintaining by vegetative propagation. In this way clones are created which are identical to the parent plant and which will therefore display any desirable characteristics. Although plants raised from seed display the visual characteristics of the mother plant they generally do not inherit any of the pests or diseases that can be transferred from mother plant to cutting during vegetative propagation.

Alpine plants often set viable (meaning good and likely to germinate) seed in the garden, some of which inevitably will grow without any help from the gardener. This seed is particularly likely to germinate if the seed has fallen amongst a layer of chippings as this helps to protect the seed from drying out in the wind or sun, or from being blown away or picked up by birds.

Seed compost and sowing

Just like compost for established plants, seedlings also enjoy a well-drained medium which encourages good root development. The compost should also be free of lumps as this will make separating a pot full of seedlings very much easier when the time comes to pot them up. Ready-made seed compost can be purchased and

Glaucidium palmatum has seeds large enough to be carefully spaced by hand.

is quite suitable though can be improved for alpine seed by the addition of around 20 per cent fine grit, perlite or vermiculite, which will improve drainage and aid separation at pricking out stage. For those wishing to make their own composts then remember that all constituents should be free from weed seeds and pests or diseases, which generally means some form of sterilization. Seed trays or pots can be used, the size of which is determined by the quantity of seed to be sown. A rough guide is that a 7cm (3in) pot can take up to approximately 25 small to medium seeds; fewer if they are larger seeds as these produce larger seedlings which will soon become overcrowded. A standard seed tray will accommodate up to 300 seeds but remember that the more you put in, the sooner they will need to be pricked out, and crowded seedlings can be very fiddly to handle!

Fill a pot or tray loosely with compost to the rim, level off and give it a gentle tap or two. Next gently firm the surface with a flat piece of wood or similar item as this helps the seeds to 'bounce' on landing and results in a more even sowing. If there are only a few seeds then these can be spaced individually between finger and thumb. Larger amounts are best sown by placing them on the palm of the hand and gently tapping that hand with the other to give a controlled fall of the seed from the inner edge of the palm – it sounds tricky but with a little practice will be found to be most effective in achieving an even distribution. It's worth practising the action with dry sand onto an empty seed tray first. Very small seeds can be mixed with a little dry sand first as this helps to gauge the evenness of sowing.

Covering seed

Pots or trays of sown seeds are then covered with fine grit, as this helps prevent them from drying out which is generally fatal once the germination process has begun. It also discourages moss, liverwort, and algae from becoming established and subsequently overwhelming the seedlings, and it allows some light to reach the seed, which can be helpful for many alpine seeds. Fortunately there is a convenient rule of thumb regarding what depth to which seed should be covered – this is that the seed should be covered by no more than its own thickness. This works well in most cases though there are as usual exceptions to the rule; some seeds for instance are flattened and these are covered by the equivalent of their thinnest dimension. Very tiny seeds are easily buried too deeply and as a result may fail to appear. Such tiny seeds are best sown directly onto a thin layer of fine grit where they will fall between the grit particles.

Once seed has been sown, remember to put a label in the tray or pot with the date and name. After many years and thousands of sowings I find that a medium hardness pencil and a white plastic label work best and last for several years. Several plants on the nursery are at least twenty years old and still have the original pencil on plastic labels. Many of the so-called indelible marker pens fade within twelve months or so and leave the gardener guessing as to what has germinated.

The best time to sow?

As a general rule seed is best sown as soon as it is ripe, because in the wild, sowing occurs once the seed has ripened and has reached the ground. Viability of seed (its ability to germinate) diminishes with age especially if kept under warm, damp conditions and therefore sowing is best carried out sooner rather than later. If you have collected seeds from your own garden this will probably be in late spring or summer and sowing some of them fresh may result in immediate germination, enabling seedlings to become well established before the onset of winter. If they do not germinate after a few weeks then all is not lost in that they will most likely appear in the following spring provided that you have kept them moist and not allowed moss and/or weeds to cover the surface of the compost.

It is safest not to sow all the seeds of one species at once; better to sow some of it immediately and save a batch to be sown during winter. Some species which require immediate sowing however are members of the *Ranunculaceae* family, for example the genus *Hepatica*, which unless sown within a few days of falling from the plant and kept moist at all times until the following spring will either produce very few seedlings or none, although germination may occur after another twelve months.

If seeds have been acquired through a catalogue or via a seed exchange scheme such as that operated by one of the many gardening societies then they will usually be received some time during the winter and should be sown a soon as possible. This will allow those which require a period of cold weather before germination can occur to receive it, and those which do not will in no way be harmed, both types germinating once warmer weather comes along in the spring.

Aftercare

Once seeds have been sown and labelled they can be placed in either a cold frame or unheated greenhouse preferably out of direct sun, as this can desiccate germinating seedlings with fatal consequences. If neither facility is available then simply placing the pots or trays outdoors in the shade will still provide conditions suitable for germination. Assuming the seeds are viable in the first place the main contributor to failure is drying out of the compost particularly at the point of germination. This can be lessened by keeping them in light shade and out of drying winds but of course you must remember also to keep the compost moist either by leaving pots to soak for a while, or watering from above using a watering can fitted with a fine rose to avoid washing the seeds out of the pot. Another problem is that of weeds and moss out-competing seedlings or covering the surface before they have had chance to even germinate, so regular inspection and removal of any weeds or moss is the best way to avoid losing seedlings.

For most growers, the excitement of germination – a minor miracle in itself – quite simply never fails to spark enthusiasm and pleasure in the same way as roots appearing on cuttings also does. In fact I make daily pilgrimages to check on how many seedlings there are in each seed tray or pot, making a mental note of the number, and getting extra satisfaction when there are a few more the next day! It is particularly important not to let seedlings dry once they have begun to appear. Also try to avoid keeping them in a closed cold frame or on a bench in the greenhouse in full sun as this can damage them so always keep them lightly shaded for the first few weeks. Remember to take precautions against slugs and snails as they will scale any obstacle in their search for the most choice *Campanula* or *Primula* known to science and raze them to the ground in one sitting!

Pricking out and potting

Before the roots of seedlings become too tangled together they must be separated and either given individual pots or transplanted and spaced out in a seed tray. Deciding on when to carry this out depends on how congested they have become, if at all. In the fortunate circumstance of only having say half a dozen

Potful of *Pulsatilla* seedlings ready to be separated and potted individually.

seedlings in a 13cm (5in) pot then they can probably be left to make small plants before needing to be separated and potted individually. When the time comes, simply tip the pot out and carefully separate them, trying to damage as few roots as possible. Then pot each in a pot large enough to accommodate the roots, erring on the side of a pot too small rather than too large. This process will be easier to carry out if the moisture content of the compost is neither too wet or dry, for example, if watering was done the day before then it might be better to wait another day or so. After watering and labelling they should be given the protection of a cold frame or unheated greenhouse for a couple of weeks to allow them to establish.

Pricking out seedlings

Seedlings which are not large enough to be potted individually are best spaced out in a seed tray or pot until they become more established. Firstly make sure the pot or tray of seedlings is neither too wet nor too dry. Too wet and the compost tends to stick together making separation tricky; too dry and the seedlings can suffer. Start by emptying the tray of seedlings onto the potting bench – with the seedlings the right way up of course! Next, gently tease them apart, handling them if possible just by the seed leaves (the first small pair of leaves to appear upon germination), and transfer to a suitable well-drained compost allowing enough space for growth until they are large enough to be either potted individually or planted out directly in the garden. As with seedlings or young plants placed straight into pots, newly pricked out trays of small seedlings should now be placed in a shaded cold frame or cool greenhouse to establish for a few weeks, after which they can be put outside prior to planting in the garden or being potted on as specimens for the alpine house.

Growing bulbous alpines from seed

There are a great many alpines that grow from bulbs, corms, tubers or some other form of storage system. Some of these reach flowering size relatively quickly: for example, some *Allium* and *Rhodohypoxis* can flower only twelve months or so from sowing. The majority of bulbs, however, require several years of growth before flowering size is attained. Genera such as *Erythronium, Trillium* and *Tulipa* may take five years or more, requiring patience in the meantime as well as careful nurturing. Quick-maturing species can be treated in the same way as non-bulbous alpines – potting as soon as they are large enough – but slower-growing types warrant a different technique that saves on space and usually cuts down on losses. Compost for sowing and the sowing itself are no different to the compost and sowing technique for non-bulbous alpines, but aftercare primarily involves leaving the seedlings together until flowering size is reached.

Bulbs benefit greatly from feeding; this is best done by applying a fortnightly half-strength liquid feed while any foliage is still green and healthy. This also applies to mature bulbs. Also once the pot has become full of roots, the whole pot of seedlings (and subsequently young plants) is knocked out and re-potted into a slightly larger size; this is repeated as necessary until flowering size is reached. If seedlings are very crowded then split the contents of the pot into two or four pieces before repotting. Always avoid potting into too large a pot as this can lead to overwatering and subsequent rotting. For example a 10cm (4in) pot would be the next size up after a 6cm (2.5in) or 7cm (3in) – remember they can always be potted up again.

A single *Sisyrinchium* plant being divided to make two new plants.

Division

This is the simplest way to increase alpine plants and can be carried out without any special facilities. Even the time of year is not as important when compared to other methods of propagation. It can be carried out on most mat-forming and creeping alpines which when dug up and separated into pieces have roots already at some stage of development. Typical examples are *Antennaria*, many types of *Gentiana*, and *Sisyrinchium*. A suitable plant can either be dug up in its entirety and rooted pieces detached and potted up for a while to establish, or the whole plant broken into two or three pieces and replanted directly, remembering to give it a good watering afterwards. Alternatively a plant can be inspected for suitable pieces with roots, without digging it up. If divisions have only a few roots then some protection from drying out in the form of perhaps horticultural fleece may be necessary for a while. The best times for carrying out division are early spring before new growth commences, or after flowering and active growth has ceased. Late summer often gives good results with enough time for the plants to become established before the onset of winter.

Cuttings

A cutting can be made from almost any part of a plant, for example the stem, offsets, root or leaf, though generally any given type of plant cannot be propagated by all such methods. Stem cuttings in themselves are divided into different types according to their individual stages of growth as follows: soft, semi-ripe, heel and hardwood.

Compost for cuttings

There are many suitable materials from which to make cutting compost but they should all result in a compost which is free from pests and diseases, drains freely, and allows roots to penetrate easily by having an open structure. A combination of any or all of the following ingredients may be used: peat or peat substitute, which is best if it has been passed through a 10mm (0.5in) sieve; sharp or river sand

The photograph shows cutting compost ingredients (clockwise from top left): perlite, vermiculite, sifted peat/peat substitute and sharp sand.

(definitely not builders' sand as it is too fine); perlite of a medium grade; or vermiculite, medium grade. These last two materials contain dust-like particles that are irritating if inhaled, so should have water added to the bag before use. Instructions are usually printed on the bag; if not, then add approximately 1 litre of water per 10 litres material and then leave to soak. A suitable mixture that will give good results consists of equal parts by volume of sifted peat or peat substitute, sharp sand, and medium grade perlite or vermiculite.

Four types of cutting (from left to right): soft, semi-ripe, heel, hardwood.

Cutting selection

When selecting a shoot from which to make a cutting choose if possible those without a flower bud and that are growing vigorously from a plant not in any form of stress, such as being under attack from pests and diseases or suffering from lack of water. Remember too that as soon as a shoot has been removed, it has no means of preventing itself from wilting, so put it in a polythene bag with a little water and keep out of the sun. In this way cuttings will stay fresh for a good half an hour or so until they can be inserted into compost and watered.

Soft cuttings

These are made from newly developing shoots of the current year's growth – either from the base of the plant or from the tips of longer shoots. Usually taken in spring or summer, the cutting length should incorporate around three to five nodes or pairs of leaves, and bend easily without kinking. The stem is cut with a sharp knife below a node and the lower leaves are also removed close to the stem. Put the cutting back into a polythene bag until ready for insertion. When inserting into the cutting medium a hole is made using a dibber, deep enough so as to bury only the base of the stem. This is then gently firmed, and once all have been inserted they are watered from overhead with a watering can fitted with a fine rose so as not to dislodge them. Label the tray or pot with a name and date before placing in a propagator with a temperature of 16–20°C (61–68°F). This type of cutting is the quickest of all to root, some taking just a few days before rooting starts; however, they are also the most delicate and soon perish if conditions are not suitable for rooting.

Semi-ripe cuttings

These are prepared in the same way as soft cuttings but from shoots that are a little older as well as perhaps longer with more nodes and consequently firmer. As a result they are a little more resilient and less likely to wilt. These cuttings are best kept in a propagator at around 18°C (64°F) although will still root – albeit more slowly – at cooler temperatures. Rooting takes longer than for soft cuttings, often at least an additional week or two.

Heel cuttings

This is basically a type of cutting that can be made from a plant that does not die back to the ground for winter. It resembles a semi-ripe cutting that has a sliver or heel of older stem at its base. This heel is achieved by pulling the cutting backwards from the main stem keeping the thumbnail behind it whilst doing so. The heel, whether firm green or slightly woody, then has any wispy bits trimmed off. Lower leaves are also removed cleanly with a knife as for soft cuttings. If left on and buried in the cutting medium they are liable to rot, which can spread to the rest of the cuttings. This method is often used on plants that are more difficult to root and/or those taken in late summer and autumn. This type can remain unrooted right through the winter, with root formation taking place in spring. Cuttings are placed in a propagator with a minimum temperature of around 7°C (45°F) or at least frost free. Satisfactory results can also be achieved using a cold frame or even on a bench in a greenhouse with the pot enclosed in a polythene bag.

Hardwood cuttings

A hardwood cutting is generally made from a plant with a shrubby or woody habit. It consists in simple terms of a stem of woody growth cut from the plant in autumn or winter, and of such a length so as to include several nodes (the point where leaves were attached to the stem and still are in the case of evergreens). This is then inserted into cutting compost and placed in a cold frame or unheated greenhouse. Rooting may not take place for several weeks but providing the compost is kept moist some rooting will have occurred by the spring.

Rosettes and offsets

Some plants do not make shoots that can be made into a conventional stem cutting – the shoot growing more in the form of a rosette or whorl of leaves such as in *Sempervivum* and many *Saxifraga*. This type of shoot requires a slightly different treatment from the usual stem cutting, being in many ways easier to take and root.

Sempervivum are probably the easiest of this type, so much so that if the rosette or offset is simply left attached to the mother plant, roots soon form, and the 'strawberry runner' stem which attaches it can be cut off and the offset potted up or even planted straight into the garden. *Saxifraga* is a large genus containing many excellent species for the alpine gardener and thankfully most, if not all of them, can be increased by detaching newly formed rosettes and treating them as cuttings. Take, for example, *Saxifraga* 'Tumbling Waters', a magnificent plant and always in short supply. Numerous rosettes can usually be found clustered around the base of a mature specimen. These are simply cut off close to the stem of the parent plant. With a sharp knife carefully cut off some of the lower leaves to leave a piece of stem long enough to anchor the rosette in cutting compost. These types of cutting are often best rooted in a cold frame or shaded position in a greenhouse, because a warm and humid propagator can often cause fungal problems. Any time of year is suitable but it is best to use only rosettes and offsets that are fully formed and not still growing. Many of the evergreen primula species and varieties also fall into this type of cutting. For example *Primula marginata* forms a clump of leaf-topped stems which can be cut off at the base and treated as an offset, with roots often already forming.

A *Sempervivum* 'mother' plant with offsets that can be detached to make cuttings.

Root cuttings

Root cuttings are used extensively in the propagation of herbaceous plants and some trees are also increased in this way. The same basic methods can also be used for alpines. *Morisia monanthos* is almost exclusively propagated by root cuttings, as seed is not often produced in this country and other methods are unreliable or produce only a small number of plants.

Firstly, make sure the plant is neither too wet nor too dry as this will make removal of the sections of roots more difficult. Plants can either be

A length of *Geranium* root cut into sections to make root cuttings.

dug up from the garden or knocked out of a pot, though plants from the ground always produce the best roots. If only a few roots are required then it is possible to dig around the plant and select what is required. Important points to remember are that the root is of sufficient thickness – an absolute minimum of 5mm (0.25in), and a minimum length of 30mm (1.5in) should be aimed for. It is a good idea when collecting and preparing the root cuttings to cut straight across the top of the root and use an angled cut at the base; in this way it should not be possible to put cuttings in upside down! Any wispy bits attached to the sides of the root are best removed, as these are liable to rot. A container of sufficient depth to accommodate the root length is filled with cutting mixture, with a little potting compost first at the base. This gives the new roots something to get their teeth into and is a definite boost to the newly developing plant. A seed tray is often too shallow, so use a pot. Next make a hole with a dibber and insert the root so that the top is level with the surface of the compost, remembering to keep it the right way up. Label and gently water, taking care not to saturate the compost. Now place them in a cold frame or greenhouse out of direct sun and with a piece of glass or clear plastic on top which should be turned over regularly. Keep frost-free for the best results. Alpines which can be propagated by root cuttings include *Geranium*, *Phlox*, *Morisia*, *Eryngium* and *Primula denticulata*.

Leaf cuttings

A handful of alpines can be propagated from their leaves, for example some species of *Primula* and *Sedum*. However, it is most commonly carried out on two members of the *Gesneriaceae* or streptocarpus family – *Ramonda* and *Haberlea*. Both can be propagated by division, though the number of plants achieved by this method is often small, whereas quite a few leaves can be detached from a mature plant. It is a fact that particularly good forms of *Ramonda myconi* are quite happy to stubbornly remain as a single rosette, making leaf cuttings the only form of increase possible. The one definite drawback with this method is the time involved: it often takes two years before a plant can be potted up or planted in the garden.

As with all methods of propagation there are particular points that are worth paying special attention to. In this case it comes during the removal of the leaf, which must be detached in its entirety as the potential for the development of a new plant is contained at the base of the petiole just where it is attached to the parent plant. A leaf removed without the full length of petiole often roots but makes no plant. Choose also larger leaves in good condition. The leaf is best removed by holding it firmly at the base of the petiole where it joins the parent plant and pulling slowly to one side – which is easier said than done! The petiole of the leaf is then inserted to the level of the lamina (leafy part) into cutting compost with a little potting compost at the base. After labelling and watering, place in a propagator with a plastic cover and a temperature of around 18°C (64°F), cooler if necessary, though rooting and plant development will then take longer to achieve. Twelve months are often required before plants can be potted. It is best to leave the old leaf on for as long as it will stay there, removing it only when it falls away at the slightest touch.

Maintenance and aftercare of cuttings

It is important to inspect cuttings on a regular basis, daily if possible, in order to maintain optimum conditions for rooting to take place. Any cuttings starting to go mouldy must be removed as soon as possible before others are affected, and spraying with a suitable fungicide may help. The cutting mixture should be kept moist but not soggy, with watering being carried out using a watering can fitted with a fine rose so as not to dislodge any cuttings. Try not to let temperatures rise above 27°C (81°F), using shading if necessary, as described earlier. Lastly, try to avoid the temptation to keep pulling cuttings out to see if they have rooted as this will quite probably pull any new roots off! Inspect them by very carefully lifting them out of the compost using a dibber or kitchen fork.

A cutting with plenty of roots, which is ready to pot.

Sooner or later the cuttings will have rooted sufficiently so that when one is carefully lifted there are several roots clustered at the base of the stem, ideally of 2–3cm (1–1.5in) length or more. Remove the tray or pot from the propagator and place in a position of good light but not direct sunlight, for example on the greenhouse bench or a cold frame with the frame light lifted up. This is to allow the cuttings to become acclimatized to a less humid atmosphere. If they have been in a polythene bag or a propagator with a lid fitted then simply open the bag or remove the lid. Should they wilt at this point then try the process in the evening when it is cooler. After about a week or so the cuttings should have got used to being in fresh air and can be potted up into individual pots, taking care to avoid damaging them in the process.

Layering

A layered shoot is one that has formed roots along its length. Layering is a method of propagation generally carried out on woody or semi-woody plants which are difficult and slow to root, though many plants layer themselves naturally and it is a simple matter to cut them from the parent plant and either pot them up or transplant directly to new homes. The method involves selecting a suitable stem, which by necessity needs to be close to or on the ground; layering is therefore suitable only for prostrate or semi-prostrate plants. The stem is then scraped with a knife, placed on the ground along with some sharp sand and a piece of stone or brick placed on top to keep it in place and help prevent drying out. This method usually allows only a few plants to be increased and takes upwards of twelve months before plants can be separated and transplanted but is useful for plants such as *Daphne blagayana*, a prostrate species with sweetly scented cream flowers, which is tricky to propagate by other means.

Plant propagation is a lot of fun and can be made as easy or as challenging as you wish. Even with the most basic of facilities there is not one month of the year in which some alpines cannot be propagated, whether by seed, division or cutting. So go ahead and give it a try – it will become addictive.

6 Pests and diseases

KEEP PROBLEMS IN PERSPECTIVE

The good news is that alpine plants are no more likely to be attacked by pests and diseases than any other type of plant. This is not surprising in such a geographically widespread and botanically diverse group of plants. Some are, as one would expect, more susceptible to attack than others. Whereas some pests and diseases can quickly become a major problem and should be controlled early on, a little slug damage or the occasional tuft of botrytis is to be expected in any garden and is pretty much impossible to avoid. Precautions can be taken against some pests and diseases, but prevention is rarely complete. In most cases total loss of the plant is not necessarily inevitable, more often simply a few damaged leaves or flowers, although things do have an uncanny habit of going wrong just when a prized treasure is at its peak of perfection! Gardeners are strongly encouraged to adopt a balance when controlling pests and diseases as a great deal of harm can be done to wildlife particularly when using chemicals. Chemical control should be used with extreme caution, following all instructions to the letter, and only used as a last resort so as to minimize any possible impact on wildlife and the environment. Biological control of many pests is a safe and environmentally friendly method using natural enemies to control numbers. The various predatory insects, mites, wasps and nematodes used are available through suppliers via the Internet.

OPPOSITE PAGE:
Caterpillars of some species of moth can cause serious damage to plants.

PREVENTION IS BETTER THAN CURE

It is possible to take simple precautionary measures against a number of pests and diseases, with cleanliness playing a major role in both the garden and alpine house. The various 'stages' of many pests and diseases overwinter amongst decaying plant material, dirty pots, grubby greenhouses and the like, so try to avoid these situations occurring. It is also important during the growing season to minimize the number of hiding places for slugs, snails and other pests, and to avoid piles of dead foliage that can harbour diseases. Any damaged or infected plant material, whether caused by pest, disease or other mishap, should be removed immediately to prevent further infection.

IDENTIFICATION

In order to assess and control a particular pest or disease it is necessary to identify it beforehand. Whereas a slime trail beside a freshly 'mown' tray of campanula seedlings condemns the slug making a hasty retreat, there are lots of instances where the culprit is harder to track down. If identification is difficult and the problem is such that it needs immediate attention then a local nursery, garden centre or parks department will most likely be able to help. Just remember to contact them before turning up and please isolate the specimen in a sealed plastic bag to avoid the possibility of spread. Many horticultural organizations also offer an identification service and many useful photographs can be found on the Internet.

Here follows a list of the most commonly encountered pests and diseases along with suggested methods of control. Please note that the illustrations of pests are not to scale.

Sciarid flies can be attracted to sticky yellow traps.

PESTS

Aphids

Commonly called greenfly, aphids can actually be black, brown, yellow or green and are a familiar pest to anyone who has any interest in horticulture. Winged adults abound from mid-spring and on throughout summer, usually flying in from neighbouring trees.

Aphids quickly multiply and use their piercing and sucking mouthparts to extract water, damaging plant tissue as they do so. They will attack an enormous range of plants but choose especially those with soft and sappy new growth. If there are enough of them, aphids will cause distortion of the foliage and stems as well as depositing a sticky substance called honeydew. Their actions can easily lead to loss of plants, particularly among seedlings. However, the damage caused can be more extensive, as it allows disease to enter wounds and, as the aphids travel from plant to plant, viruses can be transferred via their sucking mouthparts. Once a plant is infected the virus cannot be controlled, and so any infected material should be disposed of by burning. It is therefore important to control aphid numbers, which thankfully can be done in a number of safe ways.

Firstly, if numbers are manageable then squash them *in situ*, or brush off with a paintbrush and then squash them. This should be the preferred method with seedlings, which are easily damaged by chemicals. Aphids should, in any case, be controlled around seedlings before numbers get a chance to increase. Sometimes aphids can be blasted off established plants using a hosepipe as they find it difficult to climb back on, or indeed spraying them with a small amount of washing up liquid in water will prove effective. If only one particular shoot of a plant is affected then that can simply be removed and disposed of. Providing

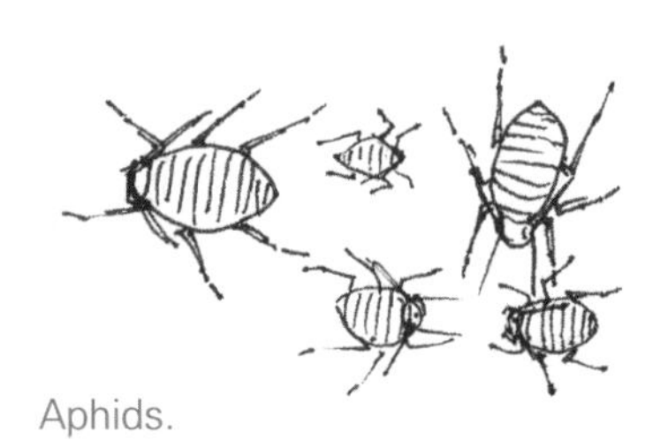

Aphids.

suitable habitats can also help to attract natural predators such as ladybirds and lacewings. Plenty of information about insect habitat homes can be found on the Internet. Alternative safe preventatives for plants grown under cover include the use of insectivorous plants or sticky yellow paper traps. Biological control in the form of parasitic wasps and midge larvae, lacewing, ladybird and hoverfly larvae can be purchased and introduced to an alpine house, with varying degrees of effectiveness based on population density and the environment within the greenhouse. They are generally most effective in warm conditions. Finally, if all else fails (which is unusual), there are many chemicals to do the job. These work either by killing on contact or by being absorbed by the plant whose sap is then taken in by the aphid.

Mealybug and root mealybug

These are pests that live and feed on plant stems, as well as roots in the case of root mealybug. Plant attack comes in the form of a soft bodied, wingless insect, around 5mm (0.25 inches) in length, which exudes a white substance like cotton wool. This woolly coating makes control difficult, especially so in the case of root mealybug as the pest lives underground and feeds on roots. Mealybugs are most often encountered on plants that have been kept in the same pots for too long. *Primulas*, *Sempervivum* and *Sedum* seem to be the first choice of alpine food plant for them. Annual re-potting of plants allows inspection of the root ball for root mealybug. Infestations can be treated with a suitable insecticide or in minor cases they can be physically removed or badly affected roots cut off and the plant re-potted in compost containing sharp sand or grit to encourage new roots to form. Ladybirds and their larvae are eager devourers of mealybugs. There is a specific ladybird 'specializing' in mealybugs, which can be purchased via the Internet, as well as a parasitic wasp.

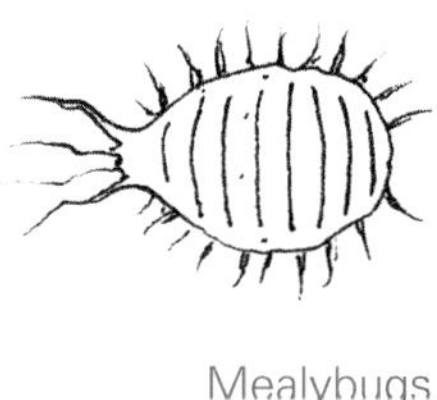

Mealybugs

Sciarid fly

Also known as fungus gnats, the larvae of these small blackish flies can be a problem in alpine houses and propagators where their small, opaque, whitish larvae, just a few millimetres long, eat the roots of seedlings, young plants and cuttings. If left unchecked they can certainly cause losses. Adult flies spend their days at or around compost level, rarely straying more than 10–20cm above it, and are more active under the warm and humid conditions of spring and summer. The larvae sometimes leave tiny slug-like trails on the compost surface and seem to prefer peat-based compost. By far the best means of control for the amateur is to maintain a layer of surface chippings on the compost, as this discourages the adult flies greatly if not completely. Sticky yellow flytraps are also effective when hung just above compost level or stuck to labels and inserted among plants. Some chemicals are available for use as a drench; however, some plants can be adversely affected by this and, as drenching also makes the compost soggy, often more harm than good can be done. Use of chippings and sticky traps, restricted use of peat and avoidance of wet compost nearly always provide adequate control.

Sciarid fly larva.

Vine weevil

The very name 'vine weevil' is often enough to make gardeners wince, and not without good reason, for this pest can inflict serious damage to many types of plants, not just alpines. As with other pests, particular genera are singled out for special attention, with

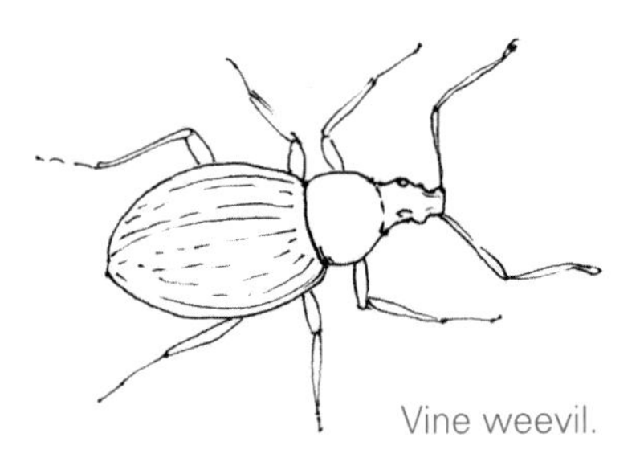
Vine weevil.

members of the *Primulaceae* and *Saxifragaceae* families being top of the list. Symptoms are not often apparent until significant damage has already been done. The plant shows no obvious signs apart from notches cut from the foliage by the adult beetle, under cover of darkness. Adults are blackish-brown in colour with a snout-like head and often lay their eggs in clusters in and around pots. The eggs are a millimetre or so in diameter and creamy-white. Both adults and eggs can be crushed easily. However, it is the plump, curved, brown-headed, cream-coloured larvae that inflict greatest damage. They chew away at the roots until every last piece has gone and the plant has nothing left to anchor it. A wilting plant sometimes arouses suspicion and further inspection of the root ball often reveals the culprit. Using a soil-based compost rather than peat-based helps deter the pest. Adults can be searched out by torchlight, and their daytime hideaways such as piles of clay pots can either be removed or employed as traps. They can sometimes be caught using sticky barrier traps made specifically for the purpose. Fortunately, there are also effective biological controls available in the form of a nematode worm, which should be administered in late summer. There is also a chemical drench that is effective. Good alpine nurseries nowadays incorporate a chemical control giving complete protection but this is unavailable to amateur growers.

Vine weevil larva.

Slugs and snails

Slug.

Both these pests are well known to gardeners along with the damage they cause. Whether large or small in size they can completely demolish plants or simply select the juiciest flower buds or fresh new shoots that would have made perfect cuttings for propagation! They also seem capable of homing in on the most treasured and rare plants in the garden or alpine house, ignoring the more common and abundant options. Slugs and snails favour warm, moist conditions and darkness and they often hide in places like the compost heap or any piles of leaf litter. Grit on alpine beds gives some slight discouragement but complete control is practically impossible. Copper tape can act as a barrier when placed around susceptible species. Biological control is available in the form of a nematode that invades slugs and snails causing bacteria to kill them. Some alpines, for example those of the *Campanulaceae* family, are particularly avidly devoured in most gardens and it is therefore best to grow them either in an alpine house, which affords a certain amount of protection, or to use slug pellets strategically placed next to hiding places or susceptible plants. Be sure to put the pellets in a space under a piece of broken pipe or stone as this should prevent them from being found by pets, birds and other harmless animals. Small trays of beer placed on the ground are also their downfall… as is a hefty stamp!

Snail.

Caterpillars

These rarely develop into a significant problem if tackled

early enough. Simply pick them off the foliage by hand as soon as they are seen. The worst types are those belonging to a group of moths that spin themselves a little web of silk, often at the tips of shoots, where they feed before moving on to new ground. The web can be opened to reveal a wriggling caterpillar, which should be picked off and squashed as it is difficult to eradicate by any other means due to the protection afforded by its silken web. This particular pest occurs most frequently under glass.

Red spider mite

These tiny reddish spider-like creatures are barely visible to the naked eye (a hand lens can provide an aid to identification). Occasionally a nuisance in warm, dry weather outdoors, it is under glass where conditions are most likely to encourage these minute pests. They tend to congregate within a barely visible web on the undersides of leaves, causing them to become mottled and yellow, eventually falling off. The best method of control is prevention – spray both plants and greenhouse during hot weather with clean water as a means of creating the humidity that the red spider mite dislikes. If possible, any susceptible plants can be stood outdoors during summer rains to discourage the pest. Chemical controls are available but it is difficult to get at the mites as the silky webbing repels liquids. A predatory mite of the genus *Phytoseiulus* predates both eggs and mites and is available from various sources via the Internet.

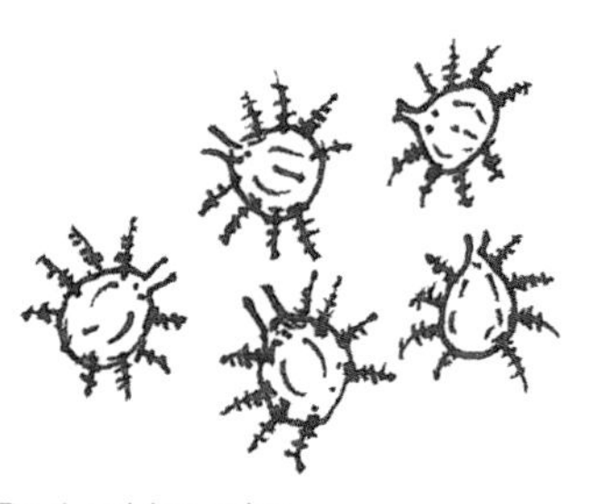

Red spider mite.

Earwigs

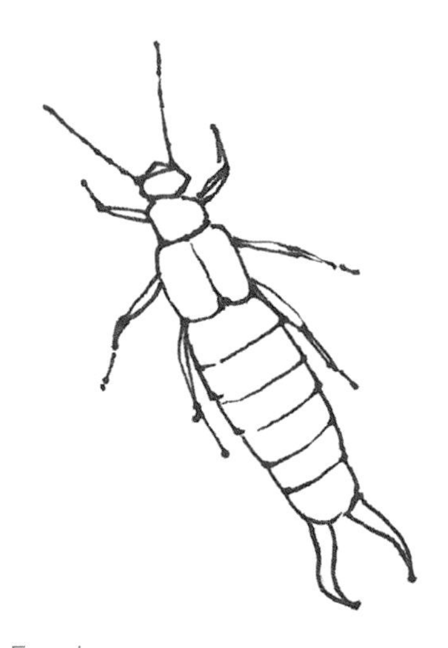

Earwig.

An easily identified insect 15mm (0.75in) in length, brownish in colour with a pair of curved pincers at the tail end. Damage caused by earwigs is unlikely to result in the loss of a plant but their night-time activities can leave young foliage shredded and, worse still, they have the most annoying habit of nibbling flowers and buds, resulting in damaged or non-existent flowers. Attack is somewhat random, with no particular pattern to their nocturnal nibbling making it difficult to predict when or where damage may occur. Fortunately an inverted flowerpot stuffed with dry grass and placed on a cane among plants provides a favoured daytime resting place from where they can be removed and destroyed or relocated.

Ants

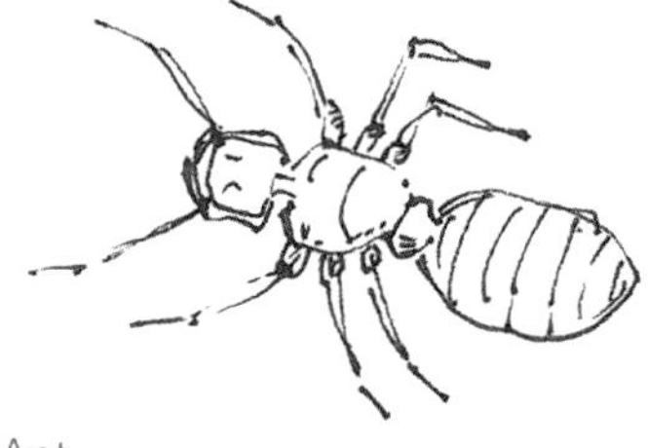

Ant.

Most gardens will contain ants, even if they are never seen. Warm, sunny gardens with sandy soils are those most likely to be affected, though it is unusual for them to be a major problem. At worst plants can be undermined, causing wilting, or the little mounds of soil that their workings produce can bury low-growing alpines. Unless a serious problem it is best simply to tolerate their presence. If intolerable then chemical controls are available.

Rodents

Mice and voles are capable of causing serious damage in the garden and especially in alpine houses and frames.

They are ever-present but can be particularly troublesome during spells of very cold weather when presumably they move under cover for more comfortable feeding. The full range of alpines is of interest to them, with seeds being top spot on the menu. A vole or mouse will, on discovering a supply of seeds, rarely leave any behind at the first sitting and if it does then it will return the next day to finish them off. Young shoots and flower buds are also high on their list. Remember that scaling a wall or woodwork is no problem for a mouse and a hole not much bigger than the thickness of a pencil is all that's required as an escape route. Various traps are available for catching rodents either dead or alive. Poisoned bait specifically for rodents, placed out of reach of birds and pets, is very effective and does not require the interaction involved with a trap.

The fuzzy fungal growth of *Botrytis* affecting a saxifrage.

Birds

Birds do occasionally pull plants apart when they are seeking nesting material or searching for insects, and sometimes apparently just for the fun of it. There is little that can be done in the garden other than placing some form of protection over any plants they are particularly attracted to. If the problem is within an alpine house, netting can be used around points of entry. Blackbirds are the usual culprits and quickly become familiar with bird scarers, eventually ignoring them. However, the odd bit of damage is a small price to pay for having such beautiful creatures around the garden.

DISEASES

The majority of diseases are encouraged by poor hygiene and insufficient ventilation, though there are some that thrive in a dry atmosphere. Diseases are most likely to be moulds, mildews or rusts and can attack a plant at any stage of growth and if left unchecked will quickly spread to others. Viruses are a group of serious diseases about which very little can be done as there is no chemical treatment. The Internet is a superb resource for detailed photographs of plant diseases and relevant forums.

Mildews

There are two commonly encountered types of mildew: powdery and downy. Powdery mildew occurs where both soil and atmosphere are dry, and can attack any type of plant. The visual indication is a powdery white fungal growth, most often on the upper sides of leaves but it can appear on the undersides too as well as flowers and stems, causing them to yellow and fall prematurely. In an alpine house, ensuring plants have adequate water during warm weather will help to avoid problems. Should any mildew occur it must be quickly controlled, to avoid rapid spread, by removing the affected parts (usually leaves) and spraying with a suitable fungicide. The fungus spores are spread by rain splash and air movement and can overwinter on dead leaves, so these should always be tidied away before spring.

Downy mildew is less often encountered among alpines but it too can attack all types of alpines. It has similar symptoms to downy mildew, though the white fungal growth is more or less restricted to the undersides of the leaves with the upper surface becoming blotched yellowish-brown, and it results in stunted growth. Humid conditions encourage this disease so plants should be given plenty of ventilation and space for air to circulate, and try to avoid overhead watering. Infected material should be removed and destroyed and plants can be sprayed with a suitable fungicide.

Rust

Rusts are not a particularly common problem amongst alpines but can cause pale, discoloured foliage along with unsightly orange-yellow pustules or patches on leaves. Usually conditions of high humidity encourage this disease, which spreads by splashing and air movement. Removal of infected parts, a drier atmosphere and adequate air circulation should help improve the problem. If not, then a suitable fungicide can be used. Some rusts are known to have certain weeds as host plants – a good reason to keep weed numbers to a minimum.

Botrytis

Botrytis, also known as grey mould, is the most common disease encountered among alpines under alpine house conditions, though is rarely seen outdoors. It attacks many different types of plants, mainly in damp or humid conditions. The softer parts of plants, particularly soft leaves and fleshy stems, become infected with the fungus causing a grey furry growth to develop that can quickly spread to other parts of the plant, especially any damaged areas such as breaks or lesions in stems. If left unchecked, rapid deterioration soon leads to the plant's demise. Control is by removal of the affected parts, improving air circulation, reducing humidity and spraying with suitable fungicide. Standing plants outdoors in dry weather or positioning them next to an open window can often help affected specimens in the alpine house.

Damping-off disease

This can affect the seedlings of all types of plants and is very difficult to control. Once established, damping-off disease can wipe out whole trayfuls of seedlings in just a week or so. Symptoms are collapse of the seedling and death from rotting stem and root; a furry fungal growth sometimes appears on the compost surface. The fungus is encouraged by moist and unhygienic conditions. Employ preventative measures by using sterilized compost and clean pots, sowing seed thinly and keeping seed trays in a clean environment with good ventilation. Use clean water for watering. The occasional watering with a copper-based fungicide may help prevent an attack.

Virus

Virus symptoms generally manifest themselves in the foliage of plants, as yellow discolouration in the form of stripes and streaks or blotches, rings or mottling. They are transmitted via physical handling, aphids and other sap-sucking or root-feeding vectors, and gardening tools (for example knives used in propagation); even seed can spread viruses. There is no cure for viruses, so preventative measures must be taken, including: controlling sap-sucking insects; avoiding handling non-infected plants after handling those that have been infected; and sterilizing any tools that have been used on affected stock. All plants infected by virus should be destroyed by burning.

7 An A–Z guide to some of the best alpine plants

Deciding on which plants to grow is always a really exciting and fun part of gardening. But with such a huge range available to the alpine gardener it pays to spend some time planning, rather than dashing off to a nursery and picking what you like the look of, only to return with a selection of heat-loving treasures for a shady north-facing crevice!

The following plants have been chosen as a result of first-hand experience. All have been grown, propagated and assessed for their suitability as good garden alpines by the author at his nursery over many years. Each possesses one or several desirable characteristics, qualifying it as worthy of consideration for planting in alpine garden situations. Most will thrive given appropriate conditions – this having been a primary and necessary quality for inclusion in this list. A number of plants have been selected on account of their outstanding beauty and even though their growing requirements may be a little more challenging to meet, they will reward any effort involved. All have been grown at 900 feet (300m) above sea level in northern England where annual rainfall averages 45–65 inches (1143–1651mm) and where temperatures ranging from –16°C to +30°C (3.2–86°F) have been recorded.

For most people, the ideal requirement is a beautiful group of alpines that will give year-round interest. With careful selection, flowers, foliage and the form of the plant itself can provide colour, shape and texture for twelve months of the year. Where a number of alpine plants are to be grown together in a rock garden or raised bed they should have similar requirements with regard to soil, drainage and aspect. Once these criteria have been satisfied then selection can be based more on their visual appearance and contribution to the overall planting scheme.

OPPOSITE PAGE:
Pulsatilla grandis.

Obviously a smaller bed or sink garden will accommodate fewer plants than a large rock garden and so take some time to plan out best use of the space. Remember that some plants will creep over or hang down the edges of beds, with others making tight hummocks between rocks. Some have a more upright habit of growth, adding vertical interest. Annual types may seed around to create natural-looking drifts, whereas herbaceous ones disappear and reappear with the seasons. Longevity of flowering may be several months with some plants, others may be only a week or two. Making careful and informed choices early on, based on a plant's cultural requirements, growth habit and visual attributes, will help to achieve a display that gives enormous pleasure and satisfaction for many years.

KEY

☼ Sun
● Shade
◑ Part shade
A Acid soil*
L Limey soil*
∩ Best with cover, e.g. alpine house
▲ Needs more care
P Poisonous if eaten
SI Possible skin irritant

* Any plants requiring acid or alkaline (limey) soil are indicated as such. Where no specific indication is given it can safely be assumed that the plant will grow in any good garden soil.

Where a plant has a name in brackets it indicates a previous name by which it was known.

Award of Garden Merit

The Award of Garden Merit (AGM) is intended as an endorsement of garden worthiness for those plants possessing it. Awards are given by the Royal Horticultural Society to plants that have been trialled to assess hardiness, ease of cultivation, quality of flowers/foliage and availability. Whereas an AGM provides an indication of a plant's garden worthiness, gardeners should always take into account their own geographical location as most of the RHS trials are conducted in southern England. It should also be remembered that by no means all plants in cultivation have been assessed. Usually only the more common species and varieties of popular genera like, for example, *Geranium*, *Saxifraga* and *Primula* have been trialled and therefore many plants worthy of an AGM have never been trialled.

Aethionema 'Warley Rose' AGM ☼

An attractive dwarf shrub with numerous wiry branches clothed in small narrow leaves and neat heads of little rose-pink flowers produced throughout summer. Eventual size is approximately 15cm (6in) high and spreading to 30cm (12in) or more. Very much a sun-lover and a good choice for a well-drained position in raised beds and rock gardens. Most easily propagated by small green cuttings taken during the growing season and placed in a propagator.

Aethionema 'Warley Rose' AGM.

Allium amabile.

Allium amabile ☼ ◑

The genus *Allium* is a very large one containing some 500 or more species and characterized by having umbels of small flowers and usually a strong smell of garlic. Many are too large to be classed as alpine and a number are somewhat weedy, seeding everywhere. This species is one of the smallest, rarely more than 15cm (6in) tall and with a similar spread, making it ideal for a sink garden or container as well as rock gardens, where it will freely produce its loose umbels of soft pinkish-red flowers all summer, without seeding unduly. The foliage is that of a fine-leaved onion. Best propagated by division after flowering. A native of western China.

Allium flavum AGM ☼ ◑

Easily cultivated in well-drained soils with some sun, Allium flavum is easily worth its place in the garden. It is variable both in height and flower colour, ranging from as little as 10–15cm (4–6in) tall to a more often encountered 20–25cm (8–10in). Its flowers can be from clear yellow to paler or even cream shades; all are attractive and worthy of consideration. The flowers

Allium flavum AGM.

are arranged in a loose umbel – some erect, others pendant – and are generally produced in profusion. Easily propagated by seed or division.

Allium sikkimense.

Allium sikkimense (kansuense) ☼ ◑

This species, also from China, is a similar size to the last and rather more upright in habit, with mid-blue 'pompom'-like flowers over the summer months. Very easy to grow, ideally in a position with full sun in any good, well-drained soil where it often seeds around, though without ever becoming a nuisance. The trio of *Allium* species listed here look particularly attractive when grown together.

Anacyclus pyrethrum var. *depressus* ☼

From the Atlas Mountains of North Africa comes this most delightful member of the daisy family. Its chief attribute is its flowers, 3cm (1.5in) wide white daisies that are a wonderful red on the reverse of the petals, freely produced for several weeks from late spring onwards. The flowers are borne on short stems, radiating from the much cut, almost feathery, grey-green foliage. The whole plant measures no more than 10cm (4in) tall, with a spread of 15cm (6in) or so. It thrives best in a position of full sun with very well-drained soil. A wall top suits it well and it is just about small enough for a sink garden, where it would gently hang over the edge. Easily propagated by seed.

Anacyclus pyrethrum var. *depressus.*

Androsace carnea subsp. *brigantiaca* ☼ ◑

Androsace is a genus of true alpine plants and contains many very beautiful species. Some can be a challenge to grow and many need to be grown in an alpine house. There are, however, several equally attractive species that are easily grown in a sharply drained spot. Like many in the genus, *Androsace carnea* subsp. *brigantiaca* grows as a compact hummock about 4cm (1.75in) tall × 6cm (2.5in) spread, with 10cm (4in) stems of white, yellow-eyed flowers in the form of an umbel. These are produced in early spring. A good choice for a sink or container.

Androsace carnea subsp. *laggeri* AGM ☼ ◑

This subspecies of *Androsace carnea* is, like the last, a native of the mountains of Europe where it can be found growing on bare patches of rocky ground amongst short grasses and other alpine plants. It is also similar in stature, making a rather spiny cushion with short umbels of pink flowers which are, in the UK, produced in the springtime. Propagation is best by taking cuttings of the shoots in autumn.

Androsace carnea sbsp. *laggeri* AGM.

Anemone nemorosa.

Anemone trullifolia.

Anemone nemorosa ● ◑

Commonly known as wood anemones and having a rhizomatous creeping rootstock with flowers on 10–15cm (4–6in) stems in spring, the numerous named forms are easy and beautiful plants for lightly shaded areas among deciduous trees and small shrubs, along with rock garden sites. The leaves are in neatly divided leaflets forming a carpet 10cm (4in) in height. Flower colour in the wild is almost always white, the reverse sometimes stained with pink or mauve. There are however many named cultivars in other colours such as 'Bowles's Purple' (a lavender blue with purple reverse), 'Royal Blue' (a good blue), 'Cedric's Pink' (pink), plus a number of doubles including 'Blue Eyes' (white with a blue centre) and 'Vestal' (with tight double white flowers, one of the very best). All are increased from division of the rhizomes in late summer.

Anemone trullifolia ● ◑

From the Himalayas this beautiful anemone has buttercup-like 2cm (0.75in) wide flowers in shades of blue, borne on the tips of wiry stems up to 10cm (4in) long, from April to June. Foliage is tidy and semi-evergreen. Dislikes hot dry conditions, so a lightly shaded spot on the rock garden or perhaps in a container is best. Propagate from division. Seed can be used but this takes much longer and requires more care.

Antennaria dioica ☼ ◑

A very tough alpine which can be grown in a variety of situations, including between paving slabs, and is vigorous enough to be used as ground cover in any sunny or lightly shaded spot. A native of mountainous regions in Europe and Asia, this summer-flowering plant has a creeping habit, rooting as it goes along, with upright 10cm (4in) stems of flowers resembling tiny pink paintbrushes. Can be easily propagated by dividing the clumps during the growing season.

Antennaria dioica.

Anthyllis vulneraria var. *coccinea.*

Anthyllis vulneraria var. *coccinea* ☼

Red kidney vetch, though not restricted to mountainous locations and more often associated with the seaside, makes a valuable addition to the alpine garden. With a mildly self-seeding habit it is great for colonizing a sunny rock garden, wall top or paved area. Flowers late spring and summer, pea-like, in clusters, bright red, occasionally orange. Height 10cm (4in) × 25cm (10in) spread. Short lived but maintains itself by seeding. A British native.

Aquilegia bertolonii AGM ☼ ◑

The flowers of alpine *Aquilegia* species closely resemble those of the taller types encountered in herbaceous borders except that they are of daintier proportions. *Aquilegia bertolonii* is especially charming, being one of the smallest species with a maximum height of around 10cm (4in) when in flower. The flowers themselves are very beautiful – a rich and intense deep blue – and the foliage is neat and not unattractive in itself. Grow it in a sunny spot with good, well-drained soil perhaps in a sink garden, container or on the rock garden where it cannot be overgrown by neighbouring plants.

Aquilegia bertolonii AGM.

Arenaria balearica ◑ ●

A very tiny alpine with minute evergreen leaves on creeping stems, which form a dense mat. The bright white flowers are equally tiny, but conspicuous, on 2cm (0.75in) stems, appearing in spring. It likes a position that does not get too dry, and will in fact happily creep over damp rocks. Best propagated by division. A native of the Balearic Isles.

Arisaema sikokianum ◑ ● P SI

Arisaema sikokianum.

From a flattened potato-like tuber arises a brownish, mottled, pointed spike, which unfurls to reveal fresh green, pointed, divided leaves, followed by a curious and very beautiful purplish-brown and white hood-like 'flower'. A striking plant best grown in deep, humus-enriched soil in a lightly shaded spot with the tuber planted 10–15cm (4–6in) deep. Propagation from natural (slow) division of the tubers or seed. A plant native to Japan.

Armeria juniperifolia AGM ☼

Like a miniature 'sea pink' this plant is worth a place in any alpine collection. It is very easy to grow in a sunny, well-drained position, and particularly well suited to sinks and containers. Forms a tight hummock of needle-like, dark green leaves with dense umbels of bright pink flowers in spring, and often with a second sprinkling in summer. Propagation is by green cuttings with a little bit of brown stem at the base. A native of Spain.

Armeria juniperifolia AGM.

Aster coloradoensis ☼

The genus *Aster* is a huge one and over the years I've grown several of those small enough to be classed as alpines. Many are rather weedy and a favourite of slugs and snails. This particular one is my favourite, with neatly toothed, greyish-green foliage forming a neat mat, above which, on short stems, are produced very attractive buds that open to lovely sugar pink daisies, all summer long. Suited to a sunny, well-drained position. 10cm (4in) tall × 15cm (6in) spread. Native to the Rocky Mountains of North America.

Aster coloradoensis

Astilbe glaberrima ● ◑

A charming little *Astilbe* small enough to be planted in sink gardens and containers with other moisture-loving alpines in light shade, where the tufts of crisply cut, shiny dark green leaves provide an attractive foil for the plumes of tiny pinkish dark-stalked flowers from mid- to late summer. The foliage often assumes nice autumn tints. The tiny seed poses problems for germination, making it therefore more practicable to propagate by division. In height about 6–8cm (2.5–3.5in), the flower spike reaching to 10–12cm (4–5in). *Astilbe glaberrima* var. *saxatilis* AGM is an even more compact version. Native to Japan.

Astilbe glaberrima.

Aubrieta 'Doctor Mules Variegata' ☼ ◑

The humble *Aubrieta* is often underrated and dismissed as too common for the alpine enthusiast. However, there are few alpines that can match it for spectacle and length of flowering. Very well suited to hanging down a wall or bank it performs best in full sun, where it can flower from March to July. It will also grow in part shade. Around 15cm (6in) tall × 45cm+ (18in+) spread. Cut back hard now and then to maintain tidiness. This particular variety has purple flowers with the bonus of attractively variegated foliage. All named forms of *Aubrieta* must be propagated by cuttings and these root well as long stemmed woody lengths taken in autumn and inserted directly into pots. Good in limey soils.

Aubrieta 'Swan Red' ☼ ◑

A deep red variety, also with variegated foliage and all the attributes of the preceding variety.

Calceolaria tenella ◑ ● A ▲

A miniature creeping *Calceolaria* from Chile. The stems are clothed in small, rounded pale

Aubrieta 'Swan Red'.

Calceolaria tenella.

green leaves, eventually making ground cover on a small scale. Happiest in light shade, with ample moisture, and will creep over rocks in such conditions. The 5–6cm (2–2.5in) flower stems rise up above the foliage with several pouch-like, pale yellow, crimson speckled flowers. Small enough for a sink or container, or perhaps a cool shaded place in an alpine house. Can be propagated by division or cuttings.

Calceolaria uniflora var. *darwinii.*

Calceolaria uniflora* var. *darwinii ☼ ◑ ∩ ▲

From the challenging climate of Patagonia comes one of the most curious of alpine plants. It grows into a 4cm (1.75in) tall cushion of rounded evergreen slightly sticky foliage, with flowers on 4cm (1.75in) stems, pendant-like 2cm (0.75in) 'clogs' in brownish-orange with a broad white stripe across the middle! It makes a wonderful display in a position with protection from hot sun and with good drainage. Well suited to sink gardens and containers, also alpine house. Cuttings are the easiest means of propagation and are best taken soon after flowering in early summer.

***Campanula betulifolia* AGM** ☼ ◑

This is one of my favourite *Campanula* species on account of its graceful beauty and habit. It makes a non-creeping, gently trailing mound about 25cm (10in) across. The 3cm (1.5in) long bell-shaped flowers appear in summer, creamy-white occasionally flushed pink. The leaves are (as the name suggests) birch-like and the stems often have a pinkish tinge. Easily grown on a rock garden perhaps where it can hang down a rock face or in a container where it could hang over the edge. A native of Turkey where it grows on rock faces. Propagated by seed.

Campanula garganica 'Dickson's Gold' ☼

The soft blue star-like flowers which appear in summer are a bonus to the evergreen foliage which is of a wonderful light yellow colour, at times darker, and makes a tidy mat 10cm (4in) tall × 20cm (8in) spread. Non-invasive and easily grown in a sunny, well-drained position in the garden or container. Can be propagated either by cuttings or by the division of a larger plant.

Campanula garganica 'Dickson's Gold'.

Campanula pulla ☼

A European species which though of a creeping habit is well-behaved enough for the alpine garden, where it will gently spread into a carpet 40cm (16in) across, made up of small fresh green leaves only a couple of centimetres or so tall. The flowers are a deep shiny purple, bell-shaped on 6cm (2.5in) wiry stems in early summer. There is also an equally lovely gleaming white form. Good for any sunny, well-drained spot. Propagation is best by division during the growing season.

Campanula raineri AGM ☼ ◑ ∩

A neat species from the Alps of Europe. Foliage is slightly downy and grey-green forming a loose cushion of 6cm (2.25in) height × 15cm (6in) or so spread. The flowers are produced during summer and are particularly large for the size of the plant, being of goblet shape 3cm (1.5in) × 2cm (0.75in) and slate-blue coloured on short stems. It likes a sunny position either on a rock garden or in a container, with excellent drainage. Seed or cuttings offer the best means of increase.

Campanula raineri AGM.

Cardamine pratensis 'Flore Pleno' ☼ ◑

This is a lovely form of the British native 'cuckoo flower' with semi-double soft lilac flowers on stems up to 30cm (12in) tall in spring. Easily grown in any reasonable soil in sun or part shade, it can also be grown among grass that is not too coarse, as in its natural meadow habitat. Propagation is by division.

Celmisia ramulosa v. tuberculata ☼ A

A beautiful plant with foliage that has the appearance of having been dipped in silvery grey ash. The leaves are small, about 1cm (0.5in) long, and densely packed around short branching stems. The whole plant takes several years to reach a height of 25cm (10in) × 40cm (16in) spread. Flowers are of the standard *Celmisia* white daisy type. Acid soil and a sunny site suit it best and a position on the rock garden or scree will allow it to slowly grow into a wonderful plant.

Celmisia semicordata 'Slack Top Hybrids' ☼ ◑ A

Probably the most spectacular plant in our garden during June when the chunky 5cm (2in) wide, white daisy-like flowers stand tall and proud above the striking sword-like silvery-grey leaves. These hybrids are descendants of seed raised plants originally collected in New Zealand which we now carefully select from our own stock. They can make sizeable plants with clumps of rosettes formed of the silver-grey leaves up to half a metre (20in)

Celmisia semicordata 'Slack Top Hybrids'.

across after several years, provided they are grown in well-drained acid to neutral soil with some sun. It is suitable for the rock garden or scree, and also makes a striking specimen for a container. Can be propagated by division of the rosettes.

Centaurium scilloides ☼

A delightful little plant with a long flowering season from late spring to autumn, when short branched stems bear clear pink flowers like little goblets with yellow anthers. A good choice for a sink or container where it will perpetuate itself by mildly self-sowing (as each individual plant only lives a couple of years at most). 4–6cm (1.75–2.5in) tall × 6cm (2.5in) spread. Found in Europe including the British Isles.

Centaurium scilloides.

Chionohebe pulvinaris ☼ ▲

Once described by one of our customers as looking like a little green mouse! It is a cushion plant from the mountains of New Zealand growing just 2cm (0.75in) or so high and slowly forming a dense silky green cushion 15cm (6in) or so across after several years. The flowers are small and white but with striking violet anthers. Being a small plant it is perhaps best suited to a sink or container where, though while never being a showstopper, it will always deserve its place.

Chrysosplenium davidianum ● ◑ ☼

A marvellous ground cover plant which happily grows in sun or shade provided the soil does not become too dry. The curious and attractive flowers, which are small and yellow, are borne in a corymb around 10cm (4in) in height, and are surrounded by conspicuous green bracts in May. Native to China.

Chrysosplenium davidianum.

***Colchicum agrippinum* AGM** ☼ ◑

Colchicum, or autumn crocuses as they are often known, make wonderful late summer-flowering additions to an alpine garden. People do complain that their leaves are coarse and untidy, but really they are simply bold, and only untidy when dying down. *Colchicum agrippinum* actually has most attractive leaves, being only 15cm (6in) in length and with a lovely wavy margin. The leaves are maintained throughout the winter and spring, dying down in summer. Its flowers are of modest proportions, approximating to those of a 'normal' crocus, purplish-pink in a tessellated pattern and very pretty.

***Corydalis* 'Kingfisher'** ◑

This splendid plant performs really well if given the conditions it likes, which are a moisture-retentive and humus-rich soil in partial shade with freedom from cold, drying winds. It has the commendable habit of flowering from spring until autumn, pretty much non-stop. The flowers are very attractive, being a lovely turquoise-blue with blackish marks, produced in profusion on a gently arching stem, and they're fragrant! Foliage is neat and forms a compact

Corydalis 'Kingfisher'.

mound some 20–30cm (8–12in) tall × 30–40cm (12–16in) spread. Perfect for growing in a container.

Cotoneaster microphyllus **var.** ***cochleatus*** ☼ ◑

A miniaturized evergreen *Cotoneaster* whose prostrate woody stems mould themselves around rocks, in time making a dense carpet. Leaves are small, dark green with a reddish petiole. Its flowers are freely produced on mature plants in a sunny position, white like those of hawthorn – but think yourself fortunate should any berries appear. Very hardy and easily grown it will eventually grow to a metre (40in) wide but just 2–3cm (1–1.5in) high. It will grow in shade but is happier with full sun. Native to China. Propagated by cuttings.

Cotoneaster microphyllus var. *cochleatus*.

Crepis incana **AGM** ☼

Pink dandelion is the common name of this delightful summer-flowering plant from the mountains of Greece. Its flowers do indeed look like pink dandelions but are borne on wiry branching stems up to 30cm (12in) tall and wide above the grey-green, somewhat dandelion-like foliage. It is a most showy plant for a sunny spot and very hardy; what's more it does not seed around like a dandelion! Method of increase is by root cuttings.

Crepis incana AGM.

Crocus ☼ ◑

A large family of dwarf bulbs, some of which are suitable for outdoors. Others require cultivation in an alpine house or frame in order to give the dormant corms a dry period. *Crocus chrysanthus* has given rise to many hybrids in a range of colours through shades of yellow, bluish and white often shaded or striped on the outside in contrasting shades. They grow well in sunny, well-drained rock gardens and raised beds, flowering in early spring. *Crocus tommasinianus* (pictured) is perhaps the most easily grown, increasing well by natural division or seed if produced. Flowers appear in early spring and vary in colour from white through

Crocus tommasinianus.

lavender to purplish-red with conspicuous orange-yellow anthers and style. The neat and narrow leaves have a distinct whitish midrib.

Cyananthus microphyllus AGM ☼ ◑

Although this plant is a member of the *Campanulaceae* it bears little resemblance except in having blue flowers. The flowers themselves are five petalled, a lovely clear blue and appear at the tips of prostrate wiry stems which are densely clothed in small heather-like leaves from September throughout autumn when few other alpines are still in flower. It is just about small enough for a sink garden or container where it will hang over the edge, and is also a good choice for a scree or rock garden, looking particularly attractive in association with autumn-flowering *Cyclamen hederifolium*. Forms a prostrate mat up to 20cm (8in) across. Best propagated by the new green shoots in spring. A native of Nepal.

Cyclamen coum AGM ◑

The genus *Cyclamen* contains within its twenty or so species around half a dozen which can be grown outdoors in the British Isles; the rest are unable to stand much in the way of frost and so need the protection of a heated glasshouse. *Cyclamen coum* is a wonderful, hardy, winter-flowering species found wild from Bulgaria to the Caucasus and quite easy to grow outside in most climates. The tubers should be planted just beneath the soil surface in moist but never soggy compost containing plenty of humus, ideally in the form of leaf mould. Planting beneath deciduous shrubs or trees or somewhere that is lightly shaded in summer usually works well. The neat, rounded leaves are often attractively marked in contrasting zones of green or silver and a perfect foil for the flowers that can be white or pink to near carmine with a dark purple nose. Plants often self-sow but seeds can be sown when ripe in the summer and grown on in the same pot for a few years before planting out.

Cyclamen hederifolium AGM ◑ ●

Cyclamen hederifolium is the best known and easiest *Cyclamen* species to cultivate. It is also highly variable in leaf with contrasting zones of green and greyish silver shades, and the shape can be broadly oval to arrowhead. These attractive leaves are produced by the plant immediately after flowering in the autumn, and persist throughout winter and spring, only dying away in early summer. The beautiful pink or white flowers push up before and throughout autumn. It likes a rich and well-drained soil in

Cyclamen hederifolium AGM.

a lightly shaded position where the tuber from which it grows will grow to saucer size and can live for twenty or more years, hopefully seeding itself during that time. The seedlings can often be found nowhere near the original plant – usually the culprits are ants, which collect the large, sticky seeds and presumably drop them off en route to their nests.

Daphne cneorum ☼

A prostrate evergreen shrub with a maximum height of 15cm (6in) and the potential to spread to 40cm (16in) or more. An absolute delight when the sweetly scented bright pink flowers are produced in summer. It looks especially pleasing growing over the edge of a raised bed, even more so if at nose height. It likes a sunny spot, well-drained soil and hates dryness at the roots. Propagated by small soft green cuttings or layering. Several variations exist including a variegated form.

Daphne cneorum 'Variegata'.

Daphne retusa ☼ ◑

Only just small enough to be classed as an alpine, *Daphne retusa* has too many great attributes not to be included among a collection of rock garden plants. With such a heady fragrance its flowers are its chief asset, and their scent can be detected from several paces in the case of a large specimen. The flowers themselves are most attractive – white inside, purple on the reverse and borne in clusters at the ends of the branches. Foliage is evergreen, dark and shiny. Pea-sized red berries are produced in late summer to autumn. Growth is quite slow at around 3–4cm (1.5–1.75in) a year, which means that a young plant could be planted in a container for a few years and then carefully transplanted to the garden. Although *Daphne* is known to resent root disturbance, I once dug up and successfully moved a ten-year-old specimen which is now twenty-six years old and approximately 1m × 1m (40in × 40in). Grows best in a position with some sun. Can be raised from the berries sown in autumn.

Daphne retusa.

Delosperma basuticum ☼ ∩

Delosperma is a genus of compact, fleshy-leaved, succulent-like plants with brilliantly coloured flowers thereby giving the appearance of being more suited to an indoor cactus collection than an alpine garden (although some members of the genus may suffer in a wet and frosty winter). The genus requires a very well-drained soil, ideally 50 per cent or more grit/sharp sand especially in wetter areas. Cold is not such a problem and *D. basuticum* has survived –16°C (3.2°F) at the nursery when kept under cover. This species makes fleshy-leaved evergreen mats up to 30cm (12in) across and for several weeks in spring and summer erupts with almost fluorescent yellow, stemless flowers each about 3cm (1.5in) wide.

Delosperma basuticum.

Delosperma from Graaf Reinet.

Delosperma from 'Graaf Reinet' ☼ ∩

This splendid *Delosperma* takes its name after an area of the Drakensberg Mountains of South Africa where it was collected. Small 1cm (0.5in) fleshy leaves clothe the congested stems, eventually resulting in a plant with a striking dome-like appearance. Requires perfect drainage and does well planted in a good depth of sharp sand and chippings with a little soil, and a sunny position where it will reward with flowers from late spring until well into autumn. Can be propagated from cuttings during summer.

Dianthus alpinus AGM ☼

An absolute gem from the alpine regions of Europe, which forms a dense evergreen cushion of narrow glossy leaves no higher than 2 or 3cm (1–1.5in) and up to 15cm (6in) in diameter. The flowers are large for the size of the plant, around 3cm (1.5in) in diameter on short stems, usually mid-pink – sometimes darker and near-red – but always with a darker central zone. There is also an attractive white form. The main flowering period is late spring and early summer but also intermittently to autumn. Cultivation is easy in a sunny, well-drained spot and propagation is best from cuttings of the new shoots.

Dianthus alpinus AGM.

Dianthus microlepis ☼ ∩

This delightful species makes a good choice for a container garden as it will almost certainly never become too large. It forms a dense, evergreen dome of approximately 6cm (2.5in) tall × 10cm (4in) spread, made up of numerous rosettes of narrow leaves each of which measures no more than a centimetre or so across. It bears stemless mid- to deep pink flowers mainly in early summer but also intermittently until autumn. Soil containing plenty of grit or sharp sand along with loam and ideally some leaf mould is perfect and a position in full sun or with a little shade will give excellent results. Cuttings can be made in late summer from 2cm (0.75in) long brown-stemmed shoots. An attractive white form is also worth growing.

Draba rigida var. *imbricata* ☼ ∩

The genus *Draba* contains many species – a large proportion of which are perhaps not ideally suited for the alpine gardener, being either rather difficult to grow and requiring an alpine house, or of little aesthetic value. Some species, however, are outstanding and this is one of the very best. It requires no special treatment beyond a gritty, well-drained soil and a position in good light, where it will develop into a dense, near hemisphere-like cushion made up of tiny rosettes of mid-green leaves. In spring its wiry flower stems bear clusters of very pretty clear yellow flowers. A most attractive addition for a sink garden or crevice on the rock garden. Maximum size attained is around 12cm (5in) diameter and 10cm (4in) tall.

Draba rigida var. *imbricata*.

Dryas octopetala AGM ☼

A dwarf shrub native to Europe and North America with creeping woody stems and small oak-like leaves, green above and greyish below and usually eight-petalled (hence the name 'octopetala'). Creamy-white flowers with a central boss of yellow stamens are to be followed by feathery seed heads. This, the normal species, is quite a vigorous alpine and needs regular attention from secateurs to keep it in check on all but the larger rock gardens.

Dryas octopetala AGM.

Dryas octopetala 'Minor' AGM is in most circumstances a better choice, being more compact and smaller in all its parts. Propagate by treating rooted runners as cuttings.

Dwarf conifers ☼ ◑

The idea of dwarf conifers for the alpine garden is an attractive one, conjuring images of miniature evergreen trees living for years in complete harmony and scale with their fellow alpine plants and setting. Such a vision of happiness is perfectly achievable provided you are armed with the knowledge of which species are 'proper' dwarf conifers ... and which are those that merely spend their first few years as dwarf conifers and the next few years as sofa-sized specimens. If lumbered with a 'dwarf' conifer whose aspirations are beyond alpine dimensions then a therapeutic remedy in the form of 'cloud pruning' can be tried. The majority of the branches are 'topiarized' to assume the appearance of a gnarled and aged specimen. If this doesn't work, shred it!

Joking aside, dwarf conifers can add greatly to an alpine garden, providing not only scale and some height but also creating microhabitats as a result of the drier soil around

Dwarf conifers. Back, left to right: *Picea abies* 'Little Gem', *Abies balsamea* 'Hudsonia Group'. Front, left to right: *Chamaecyparis lawsoniana* 'Green Globe', *Picea glauca* 'Laurin' and *Juniperus communis* 'Compressa'.

their bases coupled with shade and shelter from wind, rain and sun. Dwarf conifers are undemanding as to soil type, with a good loam-based and well-drained one being the ideal. Shaded sites result in uncharacteristic growth and can lead to an open habit. This allows the weight of winter snows to break off branches, thereby creating unsightly gaps, whereas full sun produces a densely needled, robust plant. Success with propagation is variable, with some genera much easier than others, and the especially choice cultivars often being very difficult to root. Two types of cutting are used. The quickest to produce roots are firm green cuttings taken off the current season's growth during summer and placed in a propagator – these are usually not ready to be potted until the following spring. The other type is a heel cutting where a small sliver of brown-barked wood from the previous year's growth is pulled off along with the cutting, then trimmed and placed in cutting compost in the propagator or cold frame – this being most often done in autumn.

A dwarf conifer can perhaps be defined as one that makes no more than 2–3cm (0.75–1.5in) of growth per year and therefore should not grow to look out of proportion with its surroundings for many years.

Here are a few suggestions to look out for:

Abies balsamea 'Hudsonia Group' AGM

Only just a dwarf conifer, reaching a height and breadth of around 40cm (16in) in perhaps ten years or so but possesses a very pleasingly regular architectural habit of stiff stubby branches with dark green needles.

Juniperus communis 'Compressa' AGM

The perfect conifer to add a vertical accent to an alpine garden and small enough even to be planted in a sink garden for a few years, after which it can be transplanted to the rock garden. A tight erect column 30cm (12in) in height after ten years.

Picea abies 'Little Gem' AGM

In habit, a dense, somewhat irregular mound 15cm (6in) in height by 25cm (10in) spread after 8–9 years. One of the more difficult to propagate.

Picea glauca 'Alberta Globe'

Not to be confused with *Picea glauca* var. *albertiana* 'Conica', which can reach 2m (78in) tall in only a few years, 'Alberta Globe' makes a squat cone 60 × 60cm (24 × 24in) after about ten years, formed of closely packed tufts of light green needles.

Edraianthus dinaricus ☼

Edraianthus dinaricus.

Edraianthus is a small genus of cushion plants related to the campanula family, most of which are excellent rock garden plants. This particular species forms a mat made up of small needle-like grey-green leaves no more than 2–3cm (0.75–1.5in) in height and with a spread of 15cm (6in). In summer the flowers appear, violet in colour, tubular in shape and flared at the ends. It needs a sunny position in gritty soil for best results. Propagation is best by seed. Other species to look out for include *E. pumilio*, which is smaller with more silvery leaves and paler flowers, and *E. serpyllifolius*, which has small, deep green leaves that make a dense mat, and good purple flowers often on reddish stems. This last species is variable from seed – some forms have very large flowers that open wide; others show smaller flowers of rich purple.

Epilobium glabellum ☼ ◑

It may seem madness to recommend a willowherb to plant in a rock garden as this genus includes, most notably, a vigorous weed with invasive roots and seeds that blow everywhere, but this species is one of a handful possessing none of these tiresome traits. In common with most plants from New

Epilobium glabellum.

Zealand *Epilobium glabellum* has white flowers (or at least cream), produced in great profusion along 30cm (12in) stems throughout summer and into autumn. The foliage is of a fresh apple green, complementing the reddish stems and stalks. It can be cut hard back during the flowering period to produce a second, later flush of flowers. It doesn't seed around and is propagated from cuttings.

Erigeron aureus 'Canary Bird' AGM ☼

This wonderful plant has many attributes, foremost of which is an ability to flower from April to December non-stop. Even during the other three months resting buds are often visible – a promise of what is to come. The plant has a neat and tidy habit making it an excellent choice for a sink or container garden. It also propagates quite easily from small green shoots rooted as cuttings during the growing season. Don't try to grow it from the apparently abundant seed because it is sterile.

Erigeron aureus 'Canary Bird' AGM.

Erigeron karvinskianus AGM ☼ ◑

The so-called Mexican daisy is such a delightful plant wherever grown but is especially welcome when established in a wall or in between steps or paving. It also makes a wonderful display in a container on its own as it will flower from early summer non-stop until hard frosts cut it back, often still flowering as late as November or even December. Numerous shoots neatly clothed in small pointed leaves bear the daisy flowers which are white upon opening, gradually fading to a deep pink before the seeds develop and are blown away in the wind. Approximately 20cm (8in) tall × 30cm (12in) spread. Severe cold and wet winters can damage the plant but usually a few shoots remain and vigorously re-grow once warmer weather returns.

Erigeron karvinskianus AGM.

Erinus alpinus AGM ☼ ◑

There are some alpines that make themselves very much at home in an alpine garden, creating drifts by seeding themselves around into nooks and crannies thereby helping to achieve a more natural-looking rock garden or scree. Sometimes such plants get a little carried away and

Erinus alpinus AGM.

become weeds but not *Erinus alpinus* (or fairy foxglove as it is commonly known). Its 8cm (3in) spires of fragrant pinkish-purple flowers are very welcome especially when it becomes established in walls, requiring as it does very little in the way of soil for food and moisture – it seems to find enough of both among the cracks and crevices into which it seeds. A number of named forms exist, in varying shades of mauve to near red and there is also a white form.

Erodium × *kolbianum* ☼

Erodiums are closely related to geraniums and share with them the valuable habit of flowering for several months from late spring to autumn, asking to be grown in nothing more than a sunny location and well-drained soil. *Erodium* × *kolbianum*, like others of the genus, has highly attractive deeply-cut and almost frond-like foliage. The leaf colour is a greyish-silver, complementing the flowers which are a very pale rose, with the two uppermost petals having a deep purple blotch. Flowers are held on stems just above the ground-hugging mat of foliage, which can grow to 30cm (12in) or more across.

Erodium × kolbianum.

Erodium × *variabile* 'Roseum' AGM ☼

An extremely long-flowering plant, seldom being without blooms from May until October. Unfortunately it very much dislikes wet winters and severe cold spells below –10°C (14°F). In areas of less severe weather it is a first-class plant, making soft-leaved cushions a few centimetres (1in) high and spreading to 20cm (8in) with masses of mid-pink, veined flowers on short stems. Propagated by soft green cuttings. The white form is also worth growing.

Erodium × variabile 'Roseum' AGM.

Erythronium dens-canis AGM ☼ ◑

The spring-flowering erythroniums are commonly called 'dog tooth violets' on account of the shape of the canine tooth-like bulbs from which they grow. This species is one of the smaller, having flowering stems of 10cm (4in) or so, and the flowers themselves have swept-back petals, rosy-pink in colour though quite variable. Leaves are very attractively marbled green and brown and make a good show even without the beautiful flowers. Easily grown in rich, deep soil where large clumps slowly form. They are dormant from midsummer until autumn, during which time the clumps can be lifted and separated.

Erythronium dens-canis AGM.

Erythronium 'Pagoda' AGM ☼ ◑

Reaching up to 30cm (12in) tall and with bold glossy green leaves this pale yellow hybrid

Erythronium 'Pagoda' AGM.

Erythronium is an excellent hardy garden plant flowering with profusion during April and May. Clumps soon develop from the large bulbs that should be planted with their noses 15cm (6in) deep. Quite a large plant so best suited to a larger alpine bed or border where a little light shade will allow the flowers to last longer.

Ewartia planchonii ☼ ◑

An entirely prostrate creeping alpine with small rounded silver-grey evergreen leaves from equally silvery stems giving a dense mat no more than 2cm (1in) high, and rooting along the stems as it creeps. The small yellowish flowers are incidental. Best grown in well-drained but moist soil in sun or part shade. Mats can be divided to make new plants, or cuttings can be made from soft new shoots in spring or summer.

Fritillaria camschatcensis ◑

A large genus containing many exciting species to tax the more experienced grower. There are, however, a few that are more easily grown and suitable for planting outdoors such as *F. camschatcensis* with erect leafy stems reaching to 30cm (12in) or more with a modest cluster of pendant bell-shaped flowers in a very unusual deep purplish-black colour produced in late spring. The plant grows from a scaly bulb that produces many offsets and should be grown in light shade with good soil. It occurs wild in parts of Asia and North America.

Galanthus 'S. Arnott' AGM ◑

There are so many varieties of snowdrop to choose from that 'S. Arnott' will have to be the flag bearer! It is a variety that displays all that is best in a snowdrop: firstly it is of great constitution, being thoroughly hardy and vigorous; the foliage is strong and somewhat blue-green coloured; the flowers are large and nicely shaped; and the bulbs are not too difficult to acquire from nurseries. It is easy to grow in good garden soil, and a lightly shaded position will prolong the spring flowering. The bulbs can be lifted and divided immediately after flowering, re-planting them straight away.

Galanthus 'S. Arnott' AGM.

Gentiana acaulis AGM ☼

Gentians for many gardeners are the epitome of an alpine plant and an image of *Gentiana acaulis* is the one most likely to spring to mind with its trumpet-shaped rich blue flower. It is also the gentian usually encountered growing amongst the turf in mountainous regions of Europe, sometimes so abundantly as to make avoiding treading on them an effort. *Gentiana acaulis* is probably the easiest gentian to grow, given good garden soil and an open situation where it can make 30cm+ (12in) wide mats of evergreen leaves and a magnificent display of large, deep blue, trumpet-shaped flowers.

Gentiana acaulis AGM.

Gentiana farreri ☼ A

The autumn-flowering gentians all have their origins in Asia, and are characterized by somewhat tufted and needle-like foliage that

Gentiana farreri – one of the autumn-flowering types.

dies down for the winter after flowering in late summer and autumn. They bear upward-facing, trumpet-shaped flowers in practically every shade of blue or white. They can make large, spreading mats and are a glorious sight when in flower. Nearly all of them require a distinctly acidic and humus-rich soil, kept well watered particularly during the growing season. Propagation is easily carried out by division of the clumps in spring. Usual height is around 5–10cm (2–4in) with a spread of 25cm (10in).

Gentiana paradoxa ☼ ◑

Gentiana paradoxa.

An unusual, distinctive and beautiful species from the western Caucasus. It is deciduous, producing more or less upright stems to around 20cm (8in) densely clothed in narrow leaves and terminating in beautiful, large sky-blue flowers with green spots towards the throat. Especially valuable for its late flowering in August and September, it prefers a sunny position in well-drained soil.

Gentiana saxosa.

Gentiana saxosa ☼ ◑ A

A late summer- and autumn-flowering species forming an evergreen mat of deep green, glossy leaves often with bronzy tints and short stems bearing clusters of ivory-white, upward-facing flowers in abundance. Small enough to be grown in a sink garden, it is equally at home in a raised bed containing an acidic compost and preferably including leaf mould or a little peat, where it can reach a maximum size of around 15cm (6in) across and no more than 4–5cm (1.75–2in) high. Established clumps can be lifted and divided after flowering.

Gentiana septemfida AGM ☼

A good, summer-flowering species from Asia Minor and the Caucasus, this species likes a sunny position, preferring moderately dry rather than very moist soils and makes a wonderful show on a rock garden or even at the edge of a border containing plants such as *Geranium* 'Ballerina', erodiums and summer-flowering sedums. Plants can be quite slow growing but may eventually attain 40cm (16in) or more across and a height of 10–15cm (4–6in). It flowers in clusters at the ends of the radiating stems, blue with a paler and spotted throat, mid to late summer. Stem cuttings are the best means of propagation although large plants can be carefully divided; seeds are copiously produced, though the resultant plants may have poor specimens amongst them with small and poorly coloured flowers.

Gentiana verna.

Gentiana verna ☼

The 'spring gentian', as it is commonly known, has flowers of the most brilliant and beautiful blue, perfectly set off by a compact mat of glossy evergreen foliage making it irresistible to everyone who sees it! A rare British native, it grows best in loamy, well-drained soil with plenty of moisture and good light. It makes a very nice sink garden plant, never growing much more than 10cm (4in) across and 5–6cm (2–2.5in) tall in flower, though if several plants are positioned together then a magnificent display can be created.

Geranium 'Apple Blossom' ☼

Geranium 'Apple Blossom'

One of the nicest hardy geraniums for alpine gardening and definitely worth the trouble to acquire – for it is not a plant commonly encountered in nurseries. The large, rounded palest pink–near-white flowers are stencilled with darker coloured veining and appear from May until the end of September. Foliage too is attractive, forming a clump 20cm (8in) tall with a spread of 30cm (12in) or so made up of the neat silver-green leaves. A sunny site suits it best in a good well-drained soil. Cuttings taken during summer offer the best means of increase. It is thought to be a hybrid between *G. subcaulescens* and *G. argenteum*, both of which are also good garden plants.

Geranium 'Ballerina' AGM ☼

Geranium 'Ballerina' AGM.

This wonderful plant is a hybrid derived from alpine species and is definitely worth a place on any rock garden. Very hardy and long-lived, with a continuous and abundant display of pale rosy-purple flowers with darker 'stripes' from May until the end of September, it asks only for any decent soil and some sun. A rock garden or raised bed is the best place, where it will make a rounded hummock around 20cm (8in) tall × 40cm (16in) wide after a couple of years. Root cuttings are the easiest means of increase.

Geranium dalmaticum 'Bridal Bouquet' ☼ ◑

Geranium dalmaticum 'Bridal Bouquet'.

This tough little beauty was selected by the author as a sport from *Geranium dalmaticum*, which has pink flowers and is itself a good alpine. 'Bridal Bouquet' has the bonus of producing both pink and white flowers together on the same flower stem. Foliage is attractive too and assumes nice red and orange autumn tints. Its slowly creeping habit allows it to make ground cover, and it is perfect also for growing in a not too dry wall where the stems will spread between the stones. Flowering time is summer, and it can be propagated by cuttings.

Geranium farreri ☼ ◑

A very pretty species and small enough for inclusion in a sink garden, though not too small to be planted in a raised bed or sunny well-drained position on a rock garden, where the large and beautiful pale rose flowers with black anthers are produced in early summer. The leaves are also an attractive feature, having red-brown petioles. The whole plant is generally no more than 10cm (4in) in height and scarcely much more across. Propagation is by seed, which should be collected before the seed capsule explodes and catapults the seeds out of sight!

Geranium farreri.

Geranium subcaulescens 'Splendens' AGM ☼

A wonderful geranium for a sunny, well-drained spot in good soil. Like most members of this genus it is very hardy, easy and long lived. Flowers of an unusual and lovely magenta pink with a paler centre and dark anthers are produced during summer. Foliage is pale greyish-green, borne on wiry petioles, with the whole plant approximately 20cm (8in) tall and 30cm (12in) across. Can be grown from semi-ripe cuttings taken during the summer.

Geum montanum AGM ☼

Common in the Alps of Europe. this is a splendid plant for a sunny rock garden in any good garden soil, where the compact leafy clumps give rise to 3cm (1.5in) wide golden yellow, rose-like flowers in summer, followed by attractive feathery seedheads. Very hardy and long lived as well as easy to grow, it can be increased either by division or seed. Size attained is approximately 12cm (5in) in height and 30cm (12in) across.

Geum montanum AGM.

Gladiolus flanaganii ☼ ∩ ▲

There really is an alpine gladiolus! Found in the Drakensberg mountains of South Africa, it grows amongst such precipitous rocks and cliffs as to have earned itself the nickname of 'suicide lily' – presumably due to the dangers anyone wishing to get near it encounters. Growing from a typical *Gladiolus* corm, the foliage and flowering stem reach 20cm (8in) or so, bearing several large, brilliant red flowers in August and September. It makes a spectacular addition to a collection of late-flowering alpines. It likes very well-drained compost and prefers if possible to be kept on the dry side in winter whilst the corms are dormant.

Gladiolus flanaganii.

Glaucidium palmatum AGM.

Glaucidium palmatum AGM ◑ ● ▲

For a lightly shaded position in humus-enriched soil, *Glaucidium palmatum* is a real beauty. With large, four-petalled poppy-like flowers of soft lavender to lilac, held amongst or just above beautifully veined palmate leaves, it is an aristocrat amongst plants. Height can be up to 40cm (16in) × 60cm (24in) spread, so while not a small plant it is definitely worth the space. It comes from the woodlands of Japan. Propagation is by careful division or seed.

Globularia meridionalis 'Hort's Variety' ☼

Growing into a hemisphere of evergreen, shiny, deep-green leaves and bearing short stems with delightful powder blue pom-pom flowers, this *Globularia* is a real beauty; it is also easy to grow and very hardy. Small enough to be planted in a sink garden for several years it can eventually grow to 20cm (8in) or more across × 15cm (6in) high. It flowers in summer and can be propagated by either careful division or semi-ripe cuttings in late summer.

Globularia meridionalis 'Hort's Variety'.

Haberlea rhodopensis AGM ●

The perfect plant for a shady crevice on a rock garden or north-facing retaining wall, mimicking the shaded cliffs and outcrops of its native mountain home in the Balkans. It also makes a fine plant for a container in shade. From the evergreen rosettes of toothed and somewhat bristly leaves, stems bearing several lilac flowers with pretty purple markings arise in spring. It is very hardy and propagates most easily from division after flowering but also from leaf cuttings, though these will take a couple of years to make a size suitable for planting. *Haberlea rhodopensis* 'Virginalis' is a rare, white form of the species and even more beautiful with delicate yellow marks on the throat of the flower.

Haberlea rhodopensis AGM.

Hacquetia epipactis AGM.

Hacquetia epipactis AGM ☼ ◑

In a mild winter *Hacquetia epipactis* begins to open its first flowers in February and due to cold temperatures at this time they last for weeks. New flowers can still be opening in April, thereby giving a long season of interest. The flowers are actually small and yellowish in a central boss surrounded by green bracts, giving the appearance of a larger flower. At flowering time these are stemless and form a pincushion of blooms before the new leaves appear. The foliage appears after flowering and will grow to make a clump 20cm (8in) or more high by a spread of 30–40cm (12–16in). This is an extremely hardy plant that will grow in any reasonable soil. Propagation is most easily carried out by division in late summer. Native to eastern Europe.

Helianthemum ☼

There are dozens of named forms of *Helianthemum*, or rock roses as they are commonly called, and they are an ideal alpine for providing impact in the garden. Most produce an evergreen shrubby mound up to 30cm (12in) high and can spread to 45cm (18in) or more, making them perfect for clothing a sunny bank or larger areas on a rock garden. A succession of cheerful flowers is produced all summer long in every colour except blue and green. Their vigorous habit can lead to untidy growth but this is easily rectified by a severe chop back, either in spring as growth commences or late summer. This encourages vigorous new growth and maintains a tidy habit. Propagation is by cuttings in summer or autumn.

Helianthemum 'Henfield Brilliant'.

Helichrysum milfordiae AGM ☼ ∩ A

The *Helichrysum* genus is noted for the everlasting papery textured quality of its flowers and this alpine species is no exception. Attractive red and white flower buds begin to appear in late spring, opening in early summer and lasting several weeks due to their papery texture. Watch for their entertaining habit of closing immediately at any sign of rain. The foliage forms a dense cushion of silvery-green leaves that hold droplets of moisture after the rain. Easily propagated by division or soft cuttings during summer.

Helichrysum milfordiae AGM.

Hepatica nobilis AGM ◑ ● ∩

These are popular spring-flowering plants from wooded mountainous regions of the northern hemisphere. The flowers are anemone-like and in shades of blue through purples, pinks

Hepatica nobilis AGM.

and near-red to white, with some forms having attractively patterned leaves. Beautiful on a shady rock garden or in a woodland garden, they also make lovely container plants particularly where the container can be raised up from ground level to bring the flowers closer for enjoyment. Alternatively, as many enthusiasts do, keep them in a cold greenhouse, placing them under the benching once flowering is over as shady conditions are their preference, particularly during summer. They are variable in size but often up to 15cm (6in) in height and slightly more in breadth. They can be increased by careful division of the crowns in late summer or early autumn or seed sown as soon as ripe around late May, which will not germinate until the following spring. *Hepatica nobilis* v. *pyrenaica* is a lovely variety from the Pyrenees with beautifully marked leaves and white flowers that have a pink or blue flush.

Hepatica nobilis 'Rubra Plena' ◑ ● ∩

A delightful form with tightly formed completely double pink buttons, very hardy and as easily cultivated as the single forms. Propagate by carefully separating the crowns in autumn – a fiddly operation but worth the effort to secure extra plants.

Hepatica japonica and *Hepatica pubescens* ◑ ● ∩ ▲

The Japanese hepaticas and their forms are very similar in general appearance to *H. nobilis* but the flowers are very variable in colour and form including the fabled double forms which are rare, highly prized and correspondingly rather expensive. Cultivation is as for *H. nobilis* although due to their scarcity most growers tend to give them the protection of an alpine house or cold frame. Grows wild on hillsides in Japan, usually among deciduous woodland.

Hepatica japonica.

Hosta ◑

Whilst the great majority of hostas are too large to be considered suitable for the alpine garden there are a growing number of diminutive hybrids that make delightful subjects for containers and cool, shady sites on rock gardens. They are, as a whole, unfussy as to soil provided a plentiful supply of moisture is always present. Propagation of any of the named forms must be by division, as seed-raised plants are likely to vary greatly from their parents. Two excellent varieties are 'Cracker Crumbs' with yellowish leaves bordered green and only 12cm (5in) in height; and 'Blue Mouse Ears', which has beautifully shaped oval leaves of a near-blue colour and also around 12cm (5in) tall. Both have typical spikes of purple flowers during summer.

Iris innominata ☼

Early summer sees lovely blooms produced on wiry 15–20cm (6–8in) stems from a mass of narrow leaves of similar length. Plants are very variable in colour with blues, mauve and yellows. Easily cultivated in a sunny and well-drained spot where sizeable clumps can develop, bearing dozens of flowers at any one time. Increased by division of the rhizomes or seed. The plant grows wild in Oregon.

Iris reticulata AGM ☼

This well-known species, along with its many hybrids, flowers early in the year with bluish-purple flowers. Each of the downward-facing petals has a central orange band running down it. This is a small species well suited to planting in groups in a warm, sunny position on a raised bed or bank. As flowering is very early, an alpine house or frame is a good place to enjoy them where they should remain unblemished by wind, snow and frost. If grown in pots it is best to re-pot yearly in late summer after foliage has died down, having applied liquid feed to the foliage whilst in growth. In the garden, bulbs can be left for a number of years but should still be given a little fertilizer, and divided and re-planted in fresh soil as soon as flowering diminishes.

Jeffersonia dubia ● ◑

This beautiful plant is on almost everyone's wish list but has never been easy to acquire, perhaps due to the care necessary to grow a plant to a saleable size. The flowers appear before the leaves on wiry 10cm (4in) stems, a beautiful lavender-blue in the best forms, and though only lasting for a couple of days they appear in succession over a few weeks in March and April. Leaves are also attractive – being deep brown-red on emerging and turning to green once unfurled. Best planted in a lightly shaded position in good soil. Very hardy and long lived. Propagated by seed sown fresh, which germinates the following spring.

Jeffersonia dubia.

Leptinella dendyi ◑ ☼

From neat mats of brownish-green dissected leaves arise 2–5cm (1–2in) flower stems in spring, bearing curiously attractive pincushion-like flowers of pale yellow and blackish-brown. It is a good plant for a bright spot in moisture-retentive yet well-drained soil where mats can reach up to 20cm (8in) or more across. Propagation is easily achieved by either division of the mats during growth or by removing individual shoots and treating them as cuttings. It is a native of the Southern Alps of New Zealand.

Leptinella dendyi.

Leucogenes leontopodium ☼ A

This and its similar relative *Leucogenes grandiceps* are two of the finest alpine shrubs to come from the mountains of New Zealand. The foliage is a wonderful silvery colour all year round with, in summer, white 'flowers' resembling an edelweiss (in fact its common

Leucogenes leontopodium.

name is New Zealand edelweiss). Although it looks as though it would need to be grown in hot, dry conditions as many grey and silver plants often do, it in fact requires a cool, moist position in acid to neutral soil where it should make a mat 10cm (4in) or so in height and 30cm (12in) or more across. Propagation is by cuttings in summer; or rooted individual stems can often be found and potted up until established.

Lewisia columbiana 'Alba' ☼ ◑

A species with tight evergreen rosettes made up of narrow fleshy leaves make this *Lewisia* an attractive plant for a sunny well-drained crevice on a rock garden. It is also a good choice for troughs and containers where the abundant sprays of small white flowers are produced over a long period in summer. A five-year old plant may be no more than 15cm (6in) across, with the flower stems a similar length. New plants can be produced by making cuttings of the individual rosettes with a short piece of stem, or sowing the seeds that ripen during summer. The 'normal' purplish-pink species is an equally good plant. Originates from the Rocky Mountains of North America.

Lewisia columbiana 'Alba'.

Lewisia cotyledon AGM ☼ ◑ ∩ ▲

From the Rocky Mountains of North America comes an alpine possessing two great attributes: firstly plants can be had in flower from spring until early winter; and secondly the flowers come in every colour from white through yellow and orange to pink and red, the exception being blue (although some red coloured ones have a hint of it). Flowers are also borne in great profusion with a large plant having hundreds out at any one time. Permanently saturated compost is to be avoided at all times, so grow them in a very well-drained compost, preferably wedged between rocks or in a wall where they can have perfect drainage. They also make the perfect inhabitants for a strawberry or herb pot allowing them to be taken under cover for the winter and kept a little drier. Magnificent plants can be grown in a cold greenhouse or conservatory where they may eventually grow so big as to require a 25cm (10in) or larger pot. It is an evergreen species best propagated by seed although it must be propagated from cuttings to perpetuate a certain colour form.

Lewisia cotyledon AGM.

Lilium duchartrei ◑ ▲

Growing up to 1m (39in) this is quite a tall plant for an alpine garden but is such a distinctive and beautiful lily it is worth inclusion. Unusual for a lily in having a stoloniferous or creeping growth habit, the bulbs are small – but small colonies soon develop producing their ivory-white turks-cap type, crimson speckled flowers in summer, made even more beautiful by the delicious perfume they produce (which is not of the heavy sickly sort made by many of the hybrid types). Try growing it in humus-rich soil with a little shade and shelter from strong winds.

Linaria alpina.

Linaria alpina ☼

Commonly called 'alpine toadflax' and inhabiting scree slopes in the mountains of Europe, this little plant is worth a place in any alpine garden, requiring only a well-drained sunny spot where it will produce its snapdragon flowers of mauve-purple each with a pair of orange stripes on the lip from summer to autumn. Although it is a short-lived plant, once it is established, self-sown seedlings will spring up here and there.

Linum capitatum ☼

Makes colonies of evergreen stiff-leaved rosettes with sturdy flowering stems rising to 15cm (6in) and bearing terminal clusters of wonderful glowing golden flowers in summer. Grows happily in a sunny situation with well-drained soil on raised beds and rock gardens as well as larger sink gardens. May be increased from cuttings or seed. A native of Italy and the Balkans.

Linum capitatum.

***Lithodora diffusa* 'Picos'** ☼ **A**

Lithodora diffusa 'Picos'.

Lithodora diffusa is a deservedly popular plant with its profusion of bright blue flowers in late spring and summer, even continuing into winter in mild districts. However, this form from the Picos mountains of northern Spain is even more desirable, as having a more compact habit makes it more suitable for the smaller garden and the flowers are of arguably an even more intense blue. The foliage is evergreen and forms a mat of 10cm (4in) or so high × 30cm (12in) across. It needs an acidic soil for best results and good light with ample moisture.

***Meconopsis* 'Lingholm'** ◑ **A**

The Himalayan blue poppy is well known to gardeners and this form is one of the best, being fully perennial provided at flowering time there is a growing crown in addition to the flowering one, otherwise it will surely die. Growing up to 1m (40in) in height, with deep blue poppy flowers up to 15cm (6in) across, *M.* 'Lingholm' needs to be planted in rich, fertile compost with plenty of moisture and shelter from strong winds. A large plant more suited to the herbaceous border than an alpine bed, it is propagated by division of the crowns.

Meconopsis 'Lingholm'.

Meconopsis quintuplinervia **AGM ◑ A**

Unlike the previous species, this one can be grown in a shady spot on a raised bed or rock garden in humus-rich acid compost where it will make clumps of upright bristly leaves to 20cm (8in) and produce a succession of pendant lampshade-like, lavender-coloured poppies in May and June. Unquestionably perennial and hardy, the crowns can be divided after flowering.

Meconopsis quintuplinervia AGM.

Morisia monanthos **'Fred Hemingway' ☼ ∩**

An unusual little plant from Corsica and Sardinia, having much cut, leathery, deep green leaves in rosette formation and stemless clear yellow flowers in late spring and early summer. Grows well on sandy soils kept drier if possible in the winter, wedged among rocks in a container or raised bed. Also a good plant for the alpine house. Root cuttings are the best way to increase stocks, taken during summer.

Narcissus **☼**

The miniature daffodils are justifiably popular – many hybrid varieties are readily available and recommended for good, well-drained soils and containers where they will hopefully increase year on year. The truly alpine species are not often seen, which is a pity but not a surprise as they can take several years to reach flowering size from seed, resulting in quite a high price. Do not let this put you off, however, as several of these *Narcissi* if given the right conditions and patience will seed themselves around to form naturalistic drifts. Avoid very wet soils as this can lead to rotting of the bulbs, so incorporate plenty of sharp sand or grit into the planting area. If you wish to transplant the bulbs this is best carried out once the foliage has died down after flowering. In order to

Narcissus bulbocodium var. *graellsii*.

raise them from seed this should be sown in the early autumn and the seedlings grown on together in the same pot until flowering is achieved, applying a foliar feed at regular intervals whilst they are in growth and only re-potting when they become pot bound. Some of the best wild species to try are *Narcissus bulbocodium*, *N. cyclamineus*, *N. jonquilla*, *N. rupicola* and *N. triandrus*. The smaller hybrids are also worth trying and there are dozens from which to choose.

Narcissus rupicola ☼ A ∩ ▲

From rocky areas in Portugal and Spain comes this treasure of a daffodil, scarcely more than 10–15cm (4–6in) tall. It has deep, clear, yellow flowers about 2.5cm (1in) across with a very short 'trumpet'. Flowering in spring with each bloom lasting two to three weeks, the bulbs slowly increase by offsets; they are also not too difficult to raise from the seed, provided you are patient! Grows best in a slightly acidic soil, well drained and never waterlogged. It makes a very nice subject for a pot or container or perhaps at eye level on a raised bed.

Narcissus rupicola.

Olsynium douglasii AGM ☼

After a summer resting period, mid-autumn sees sharply pointed new growths show themselves above ground, though they make little perceptible progress until the lengthening days and rising temperatures of late winter draw the slender shoots upwards to 15–20cm (6–8in) or so. The first flowers can be seen in the last days of winter, glorious pendant purple bells among grassy grey-green foliage. Looking delicate and vulnerable, they are in fact very hardy and well able to withstand the often wintry weather at this time of year. Clumps develop slowly but with every passing year become increasingly more splendid and beautiful. Propagation by division best carried out once the plants have gone dormant in summer. There is also a white flowered form. Native to North America.

Olsynium douglasii AGM.

Origanum amanum AGM ☼

This alpine marjoram is very attractive with pink, leaf-like bracts at the ends of the flowering shoots, among which are the flowers, like elongated little pink trumpets, appearing from June until August or September. The foliage is aromatic as would be expected, and neat – no more than 15cm (6in) tall and spreading to 25cm (10in). Definitely a sun lover and best planted in well-drained soil. Large plants can be propagated by division in the spring or by taking soft green cuttings and placing them in a propagator.

Origanum amanum AGM.

Oxalis enneaphylla 'Patagonia' ☼

The genus *Oxalis* is very large, containing several hundred species all with similar characteristics including: somewhat funnel-shaped flowers that close up at night or in poor weather; the petals closing with a twisting action; and clover-like leaves. The genus can be divided into two types by growth habit: those with creeping thread-like runners or rhizomes, which should not be planted anywhere near treasured plants; and those that grow from a bulb-like scaly tuber thereby remaining where they were planted – it is this last group with which we shall concern ourselves. *Oxalis enneaphylla* 'Patagonia' is one of the scaly tuber types and probably a geographical variant of *O. enneaphylla* with especially large, rounded purply-pink flowers with a distinctly pleasant soapy scent. Flowering over the summer months, it is easy to grow in a sunny, well-drained position where compact clumps up to 20cm (8in) or more across and 10cm (4in) high soon develop. A native of South America and very hardy.

Oxalis enneaphylla 'Sheffield Swan' ☼

Another form of the species, this one has large ivory-white greenish-centred flowers mainly during summer but continues to produce blooms to late autumn. Foliage is an attractive grey-green, in neat leaflets, forming low mounds in any sunny and well-drained rock garden feature. It also makes a good plant for an alpine house. Easily propagated by careful division of the tubers, this is best done while dormant in late winter.

Oxalis enneaphylla 'Sheffield Swan'.

Oxalis laciniata ☼ ∩ ▲

This is a real gem among the genus though unfortunately it is rather slow growing, making it perhaps only really suitable for an alpine house or sink garden. Tiny scaly tubers give rise to the equally tiny, crinkly grey-green leaflets on 2cm stems. The flowers appear in early summer – about 2cm (0.75in) across, in varying shades of violet-blue with veining, and with a nice soapy scent. Appreciates leaf mould and sharp sand in the compost.

Oxalis laciniata.

Oxalis 'Slack Top Hybrids' ☼ ∩ ▲

A range of hybrids of the neatest habit, all having attractive foliage with flowers varying in colour and including near-white with blue veining, bi-coloured blue and pink, rich deep blue, and clear rose. Preferring a sunny position in well-drained, humus-enriched compost, these hybrids do particularly well in an alpine house where they can be kept a little drier in the winter.

Oxalis 'Slack Top Hybrids'.

Penstemon rupicola AGM ☼ ◑

A prostrate evergreen shrublet from North America around 10cm (4in) tall × 30cm (12in) spread with an abundance of dark purplish-red 3cm (1.5in) long snapdragon flowers in early summer and spasmodically thereafter until autumn. Makes a great plant for growing over the edge of a large sink garden or creeping down between the rocks of an outcrop. Unlike some penstemons this one rarely dies off in patches. An excellent plant for any alpine garden, it is hardy and long lived. Grown from seed it shows wide variation though many resultant plants are poor flowerers and so this plant should be propagated from soft green cuttings in spring or summer.

Penstemon rupicola AGM.

Phlox douglasii 'Crackerjack' AGM ☼

Varieties from the *douglasii* species of *Phlox* are, on the whole, the neatest and most compact. They form a dense mat of prostrate branches clothed in needle-like leaves and bear stemless, 2cm (0.75in) wide flowers during summer. Easy plants to grow in virtually any conditions with some sun, they are especially good at growing over and down walls and banks where mats 60cm (24in) or more wide, covered in hundreds of flowers, are easily achievable. 'Crackerjack' is a gorgeous variety with very bright red-pink flowers. Seed is rarely set and so soft green cuttings in spring or summer are the best way to propagate alpine *Phlox* cultivars. Other good forms include: 'Eva', with two-tone lavender-purple flowers; 'Apollo', which is very compact with nicely rounded, deep lilac flowers; and 'Kelly's Eye' – a more spreading habit and pale pink flowers with a darker stripe down the petals.

Phlox douglasii 'Crackerjack' AGM.

Physoplexis comosa AGM ☼ ◑ ∩ ▲

Though belonging to the Campanulaceae family the flowers of *Physoplexis comosa* in no way resemble the bells of a typical *Campanula*. They are both curious and beautiful, resembling a pincushion composed of tiny purple-tipped lilac bottles that any gardener cannot fail to be

Physoplexis comosa AGM.

fascinated with. Its chief enemies are slugs and snails, therefore *Physoplexis* is best planted in a sink garden somewhat out of their reach. Try to wedge it between rocks to echo its natural growth habit in crevices among rocks, so giving it perfect drainage. Seed is the only practicable form of increase. A native of the Italian Dolomites.

Phyteuma hemisphaericum ☼

Although a member of the Campanulaceae family, *Phyteuma hemisphaericum* bears little resemblance to one, having tight flower heads composed of small blue flowers among triangular green bracts held on wiry 10cm (4in) stalks during summer. Neat, narrow leaves form a tight tuft of foliage. From the Alps of Europe, this alpine is excellent in a sink garden or gritty and well-drained raised bed, scree or rock garden. It is best propagated by seed.

Polygala calcarea 'Lillet' AGM ☼ ◑ ▲

A quite remarkable little shrublet due to its very long flowering period from April to November. A small plant, no more than 2–3cm (0.75–1.5in)

Polygala calcarea 'Lillet' AGM.

high and perhaps 10cm (4in) spread, it bears evergreen foliage and masses of small, deep blue pea-like flowers, making it a good choice for a sink garden or container. It prefers a loamy soil with some sun and is propagated by small green cuttings.

Polygala chamaebuxus AGM ☼ ◑

This species, from the mountains of Europe, is larger and more robust than the last, growing up to 10cm (4in) high and with a mildly creeping habit. It can spread to 40cm (16in) or more, although it is quite slow growing and would take several years to attain this size. The deep green evergreen foliage is attractive and a perfect foil for the abundant pea-like, yellow-winged cream flowers in spring and intermittently throughout summer. It is very hardy and easy to grow in good soil with full sun, though part shade can be beneficial in hot regions as this is a plant that enjoys a cool and moist climate. It can be propagated either by division or cuttings.

Polygala chamaebuxus 'Rhodoptera' ☼ ◑

An equally desirable form of the species with purple-pink winged golden flowers instead of the cream and yellow found in the type species. Both have fragrant blooms.

Polygala chamaebuxus 'Rhodoptera'.

Potentilla eriocarpa ☼ ◑

Alpine potentillas are generally a hardy, easy and floriferous genus of plant and this species is one of the best, having large, rounded clear yellow flowers freely produced throughout summer and autumn on neat mats of tufted grey-green leaves. Only a couple of centimetres high (1in) and spreading to 15cm (6in) it is just small enough for a sink garden or container but equally at home on a rock garden. A native of the Himalayas, it is best propagated by the fluffy seed, though it can also be increased by cuttings or careful division.

Potentilla eriocarpa.

Potentilla nitida 'Rubra'.

Potentilla nitida 'Rubra' ☼ ◑

A delightful *Potentilla* with much-branched, woody stems clothed in pinnate silvery green leaves, forming a close mat only 2–3cm (1in) high and spreading over time to 25cm (10in) or more. The beautiful 2.5cm (1in) wide rose-pink flowers on short stalks are produced during the summer months. Easily grown in a sunny or lightly shaded spot and propagated by division or cuttings in summer, this alpine from the Italian Dolomites is very hardy and long lived and a good choice for an alpine garden.

Primula

This is such a large genus that it is difficult to know where to start. More than 500 species are known, along with hundreds more hybrids and cultivars, and a great many of them are suitable for the alpine gardener. This makes selection for recommendations appear somewhat arbitrary, though the following are good plants for the alpine gardener and not too difficult to grow or acquire.

Primula auricula AGM ◑ ●

Primula auricula AGM.

Ideal for a slightly shaded spot and perfect for a shady wall with plenty of moisture, this is one of the parents for all the gaudy 'auriculas' seen in magazines and shown on gardening programmes. *Primula auricula* is a variable plant, at its best a very beautiful plant with attractive leaves often covered in a dusty grey-white powder called farina, and clear yellow flowers with a distinct white central zone and a delightful fragrance. It is also very easy to grow and extremely hardy. Although an excellent garden plant it also makes a wonderful show when grown in a cold greenhouse. Flowers appear in spring. Division of the offsets offers the best means of increase. In the wild this species mainly occurs in alpine regions of Europe.

Primula auricula 'Blairside Yellow' ◑

A very pretty plant that makes an ideal subject for a sink garden or well-drained but moisture-retentive spot on a raised bed or rock garden, with shade from the heat of the summer sun. Slow growing but hardy and long-lived, it makes neat clumps of rosettes about 10cm (4in) high with a similar diameter that give rise to umbels of clear golden-yellow flowers in late spring. Individual shoots can sometimes be detached as rooted offsets; un-rooted shoots can be made into cuttings.

Primula auricula 'Blairside Yellow'.

Primula auricula 'Slack Top Red' ◑ ●

This is a garden hybrid with foliage of the species and flowers of good size and shape, a beautiful warm terracotta-red with a clear white disc at the centre. Overall height is 15cm (6in).

Primula auricula 'Slack Top Red'.

Primula denticulata 'Birthday Beauty' ☼ ◑

Aptly named as the 'drumstick primula', *Primula denticulata* is a tough and adaptable plant able to tolerate cold, heavy, moist soils in sun or part shade. In early spring up to 100 individual flowers make up a globe-like flower head, followed by bold clumps of foliage up to 30cm (12in) across. Plants should be purchased in flower for the best colours. Often the colour can be a wishy-washy mauve, but some nurseries sell carefully selected forms in striking red, pink or purple shades that have

Primula denticulata 'Birthday Beauty'.

been propagated by division, such as the 'Birthday' series raised by Slack Top Nurseries.

Primula marginata AGM ◑ ● ☼

A great favourite with alpine gardeners, this hardy species is easily grown in a well-drained position provided that hot and dry conditions are avoided, for like most primulas it soon wilts under the stresses of excess heat and water shortage. Some shade is preferred and they do almost as well in full shade. Under suitable conditions plants develop numerous stems, building into a sizeable clump perhaps 15cm (6in) in height × 20cm (8in) or more spread, with each stem bearing heads of purplish-lilac fragrant flowers in spring. The flowers themselves are quite variable in size, shape and colour but in many respects play second fiddle to the leaves whose surfaces are often covered in silver or golden farina and can have deeply toothed margins. There are many named forms, some of which are very lovely. Seedlings as already mentioned are variable, sometimes possessing small, poorly shaped flowers but often are every bit as good as the named varieties. *Primula marginata* is a good choice for a container garden, at least for a few years, and does well planted in a shady wall as well as in rock gardens and raised beds. The named forms must be propagated by cuttings or division, but it is fun to raise plants from seed to see what you get! In the wild it is found among the Maritime Alps.

Primula marginata AGM.

Primula × *pubescens* 'Freedom' ☼ ◑

Primula × *pubescens* occurs naturally in the wild and has given rise to a group of excellent garden plants on account of their ease of culture, free-flowering nature and beauty. Well suited to sink and rock gardens, when grown in good gritty soil they reach approximately 10–12cm (4–5in) high, form clumps 15cm (6in) or so across and are easily propagated by firm cuttings or division. 'Freedom' is a form named on account of the profusion of deep purplish-lilac flowers. It has dark green foliage and, like other '× *pubescens*' hybrids, is an easy and rewarding plant.

Primula × *pubescens* 'Slack Top Violet' ☼ ◑

An excellent hybrid carefully selected from a batch of seedlings from *Primula* × *pubescens* parents. Of very compact habit and having nicely rounded purplish flowers with a white central zone, this is a distinctive variety. Good hardiness and a floriferous nature make it an ideal

Primula × *pubescens* 'Slack Top Violet'.

choice for containers or rock garden habitats. The height of the hummocks is about 8–10cm (3–4in). Propagation is by division or cuttings; both operations are best carried out after flowering.

Primula scotica ☼ ◑ ∩ ▲

Primula scotica.

This beautiful little *Primula* has a very restricted distribution in the wild and is found only at the northernmost tip of Scotland and the Orkney Isles. The neat tufts of leaves give rise in spring, and sporadically throughout summer, to 3–6cm stems of yellow-eyed, rich purple flowers. An absolute gem – best grown in a container, sink garden or perhaps alpine house where its exquisite beauty can be enjoyed at close quarters. The seed, though very small, provides the easiest form of increase, and is best sown with only a light covering of fine chippings.

Primula sieboldii AGM ◑

A Japanese species easily cultivated in any good soil with ample moisture during growth, and preferably with a little light shade. It makes bold clumps of crinkly foliage from slowly creeping rhizomes up to 15cm (6in) high × 40cm (16in) or more spread. Flowering is from mid-spring onwards on stout stems. The most common colours are shades of pink but some forms have a decidedly bluish colour and good clean whites are not unusual. Petal shape is also very variable from entire or plain to those attractively cut into snowflake-like patterns; indeed, so many colour forms and shapes have given rise to societies devoted solely to this one plant. Propagation is easily carried out by division of the dormant crowns in late summer or autumn. Also it is great fun to raise plants from seed and pick out the best ones.

Primula vulgaris AGM ◑ ●

Primula vulgaris AGM.

Although perhaps not an alpine in the strictest sense, the British native wild primrose is such an easily grown, beautiful and long-flowering plant that it is worth finding a shaded spot in the alpine garden for it, where the first soft yellow blossoms can appear at the end of winter and continue through until late spring. They are fragrant too, if you can get close enough to smell the scent. It prefers a good rich loamy soil and will seed itself around in grass if there are some little bare patches of earth for seedlings to establish. If you are unlucky enough not to get seedlings then divide established plants after flowering.

Pritzelago alpina ☼ ◑

Formerly called *Hutchinsia* before the botanists had a good look at it, *Pritzelago alpina* is a tough cushion- or mat-forming alpine with tiny glossy deep green leaves sometimes with

Pritzelago alpina.

a purplish tint, which in spring covers itself with neat heads of gleaming white flowers for weeks on end. Particularly good as a paving plant, its seeds find suitable crevices in which to germinate and grow, quickly forming glossy evergreen cushions 5–8cm (2–3in) high and spreading to 30cm (12in) or more. It grows in any reasonable soil, preferably with some sun, and may be propagated either by seed or by detaching ready-rooted stems and planting direct into a suitable position.

Pulsatilla grandis ☼

Many gardeners are familiar with Pulsatillas or 'pasque flowers' as they are commonly known, with their large goblet-shaped flowers (though some species have smaller pendant blooms), followed by decorative feathery seed heads. Hardy and often long lived if planted in good deep soil with a sunny aspect, and each year gaining in size, they make excellent plants for all manner of rock garden features. *Pulsatilla grandis* is one of the more unusual species, with very large, rich bluish-purple flowers which emerge from buds covered in silky soft silvery-grey hairs, opening just as the leaves appear in early spring. It is a very beautiful plant and should be propagated from seed which is best sown as soon as ripe in summer; the seedlings when flowering size will show variation in size, shape and depth of colour but all will be lovely! The height in flower is approximately 15cm (6in) and the dissected herbaceous foliage makes a clump around 25 × 25cm (10 × 10in).

Pulsatilla grandis.

Pulsatilla vulgaris **AGM** ☼

A very rewarding alpine highly recommended on account of being very hardy, easy to grow and beautiful. This is the 'normal' and readily available species with deep purple flowers and a central boss of golden stamens followed by feathery seed heads and dying down for the winter. It grows best in good, well-drained deep compost perhaps on the limey side of neutral though will also tolerate more acid soils, and is capable of becoming large clumps up to 30cm (12in) or more across with dozens of flowers on 15cm (6in) stems making a wonderful sight on a sunny spring rock garden. There are numerous variants with colours ranging from white through palest pink to deep pink and a true bright red, some of which are listed below.

Pulsatilla vulgaris AGM.

Pulsatilla vulgaris **'Papageno'** ☼

A strain including all the colours above but with the petals slashed and cut to give a surprisingly attractive shaggy appearance – often they are semi-double as well. An unusual and distinctive variation that always attracts comment.

Pulsatilla vulgaris 'Papageno'.

Pulsatilla vulgaris **pink flowered** ☼

Also known as 'rosea', this variety ranges from a very pale pink through deeper shades, to a wonderful clear rose colour. Like all pulsatillas they do not like to be disturbed once planted as transplanting tends to break off the taproot with fatal consequences.

Ramonda myconi AGM ●

A distinctive genus of only three similar species, this one occurs in limestone regions of the Pyrenees generally on completely shaded and sheltered moist rocks. It forms rosettes of crinkly rounded leaves with copper-coloured hairs on the undersides. The rosettes can be 15–20cm (6–8in) across, and 15cm (6in) in height, from which in spring arise short stems bearing flowers in shades of mauve with a central cone of yellow stamens. Some forms have white or pinkish flowers. It is an extremely hardy plant when planted in a cool, moist spot such as in a north-facing wall, or wedged between shaded stones on a rock garden. The plant possesses remarkable powers of recovery should it ever become parched. Even in a shrivelled and limp condition, upon receiving a good soak it will, in the course of a day or so, plump up as if nothing untoward had ever happened. In extreme cases some of the leaf margins become crispy so do not abuse this facility! Propagation can be by division or leaf cuttings in summer, or seed can be sown but is very tiny and requires careful nurturing to reach flowering size, which often takes 3–4 years.

Ramonda myconi AGM.

Ranunculus alpestris ☼

A charming alpine buttercup from the Alps of Europe – easily grown in well-drained and gritty compost with plenty of moisture in the growing season, and ideal for a sink garden or sunny position on a raised bed. It is a small plant with lobed glossy green leaves on short petioles forming neat tufts, and 5cm (2in) stems with shining white, crinkly petalled flowers in April and May, and sporadically throughout summer. Plants may be propagated by careful division of the crowns. Treat them as cuttings, or if seed is set then sow this immediately and it should germinate the following spring.

Ranunculus alpestris.

Rhodiola rosea ☼

Rhodiola is a genus of sedum-like plants with numerous ascending stems clothed in fleshy

Rhodiola rosea.

leaves, and clusters of small flowers at their tips. *Rhodiola rosea* is a widespread and hardy species which looks particularly pleasing when grown between rocks where it will form an architectural mound of stems 30 × 30cm (12 × 12in) during summer, dying back to a knobbly crown for the winter. A position with some sun is best, in any decent soil with reasonable drainage. Propagate by division, seed or by taking non-flowering shoots in late summer and treating them as cuttings.

Rhododendron ☼ ◑ A

Rhododendron 'Snipe'.

This vast race of shrubs varying in height from a few centimetres to several metres contains a number of excellent and desirable plants among its smaller species and hybrids, making perfect dwarf shrubs for humus-rich, moist but well-drained acid soil in good light. Hot and dry conditions must be avoided, as must waterlogging of the soil, which can quickly lead to root rot. They need little maintenance though can be pruned hard to encourage basal shoots, and benefit from an annual top dressing of leaf mould and sharp sand. Propagation is best by small green cuttings in summer or heel cuttings in autumn, remembering to use ericaceous compost for potting. Seed can be used but seedlings are tiny and need careful handling, also the resultant plants will often vary from the parent plant. Some species are small enough for a sink garden whilst others make better subjects for a rock garden. Good ones to try include: *R. campylogynum* AGM with small, bell-shaped flowers in reds, pinks, purples and white; *R. camtschaticum*, a dwarf deciduous species having large, flat-faced red, pink or white flowers; *R. keleticum* with large purplish flowers; and *R. fastigiatum* with attractive blue-green leaves and mauve flowers. A great many dwarf hybrids are worth growing – among them are *R.* 'Ginny Gee' AGM (white, flushed pink), 'Oban' (very dwarf, red-pink), and 'Patty Bee' AGM (large yellow flowers), and 'Snipe' (pictured).

Rhodohypoxis baurii, deflexa and named varieties ☼ A ▲

These are dwarf bulbous plants from the Drakensberg Mountains of South Africa where they stud the turf with their five-petalled starry flowers in colours ranging from white through pinks to near red. In the garden they must never dry out whilst in growth or they will become prematurely dormant. Kept well watered, many varieties flower from spring until late summer. Winter can be a problem time as saturated ground and freezing temperatures below –10°C (14°F) lead to losses. In localities experiencing such conditions it is wise to lift and dry off the bulbs, storing them under cover before replanting in spring. Many *Rhodohypoxis* enthusiasts living in cold areas grow their plants in containers, allowing them to be more easily given winter protection. Well-drained soils in milder areas allow the bulbs to thrive in raised beds and rock gardens, forming large clumps in just a few years and providing a colourful sight throughout the summer months. As well as protection from cold wet soils they also require well-drained acid compost in a sunny situation for best results. The small bulbs or corms can be separated at any time of year but is particularly successful during the growing

Rhodohypoxis baurii hybrids.

season, remembering always to give them a good watering afterwards. Maximum height is around 10cm (4in), often less. Some of the best are listed below:

Rhodohypoxis deflexa

A delightful species more dwarf than most at only 4cm (1.75in) and multiplying rapidly to form a grass-like sward with an abundance of small, red-pink flowers.

Rhodohypoxis baurii

This species and its varieties are the ones most commonly encountered. The tufted grass-like foliage provides a perfect foil for the usually reddish-pink flowers, produced in succession throughout the summer.

Rhodohypoxis 'E.A. Bowles'

An old, well-known variety, 10cm (4in), strong growing and with large mid-pink flowers.

Rhodohypoxis 'Pictus'

A slow-growing, large white-flowered form, 10cm tall (4in), with just the slightest hint of pink at the edges and bases of the petals.

Roscoea cautleoides AGM ◑

Roscoea are members of the ginger family from central Asia with flowers reminiscent of some orchids. They like a good, rich, deep and moist soil in a lightly shaded spot for their thick tuberous roots to delve into, where they will quickly develop showy clumps. *Roscoea cautleoides* can grow up to 25cm (10in) tall with spikes of pale yellow-lipped flowers in summer.

Roscoea cautleoides AGM.

The thick roots can be carefully divided whilst dormant or new plants can be raised from seed sown in late winter; these should make nice flowering sized plants in about three years.

Salix 'Boydii' AGM ☼

An invaluable plant for providing a little height to alpine gardens, this willow has the appearance of a gnarled old specimen when only a few years old. It forms a much-branched upright miniature 'tree' with stiff grey oval leaves, whitish underneath and in spring a handful of comparatively large upright chubby catkins. In winter the structure is maintained, providing some interest until buds swell and burst in spring. Very easily grown in any reasonable soil with sun, where a few centimetres of annual growth can be expected. This slows with age, giving a plant perhaps 45cm (18in) tall and with a spread of 35cm (14in) after twenty years or so. As with most willows propagation is easy from small green cuttings in summer. The plant has an interesting history – a single wild hybrid plant was 'discovered' around a hundred years ago by a Dr William Boyd whilst he was out walking in Forfarshire, Scotland, and all plants subsequently in cultivation are the result of that chance find.

Salix reticulata AGM ☼ ◑

Another splendid willow native to Scotland and England. Completely prostrate in habit with shiny, ground-hugging stems clothed in beautiful oval deep-green leaves crinkled with veining, hence the species name. The slowly creeping habit forms a plant of around 30cm (12in) in five years or so; any shoots needing to be cut back can be used to make cuttings. The plant looks particularly pleasing growing over rocks and is easily grown in a position with some sun and plenty of moisture.

Salix reticulata AGM.

Sanguinaria canadensis 'Plena' AGM.

Sanguinaria canadensis 'Plena' AGM ◑ ●

This is the double form of 'Bloodroot', so named on account of the red sap that exudes from any broken part of the thick rhizome. The rhizomes slowly creep through humus-rich soil in lightly shaded situations producing hand-sized rounded leaves held horizontally on 15cm (6in) petioles. The flowers are exquisite. Displayed in spring before the foliage, the pure white petals form a loose globe-shaped flower which, though only lasting a few days at best, are so strikingly beautiful as to warrant making a special spot in which to plant a specimen. It is a very hardy plant, found growing wild in North America, and though rather slow to build up a sizeable clump is reliable enough once established. Propagation is by division of the rhizomes in late summer or autumn.

Saxifraga

A large genus contributing a great many alpine plants perfectly suitable for rock gardens and providing the alpine gardener with a number of species and hybrids ideal for any situation. Because there are so many saxifrages, they have been divided into sections. The most popular of these from the alpine gardener's view are the 'encrusted saxifrages' – characterized by their mat-forming growth of rosettes of stiff grey-green leaves, often with beautifully encrusted margins made up of dots of lime and giving rise to spectacular sprays of white flowers during the summer months. Another section is that of the Porophylla, or Kabschias as they used to be known (image shows a selection of Porophylla hybrids). These make tight mats of small and quite hard or bristly rosettes, blooming very early in spring with single flowers or sometimes a few-flowered umbel in white, pinks, yellows, and shades in between. The Englerias are another group notable and desirable for their beautiful foliage, which is often bluish in colour with numerous lime pits; the flowers in stiff sprays are usually reddish or purple. Other species do not fit easily into any particular group and among these are many other outstanding plants. They can all be propagated by making cuttings of either the individual or small clusters of rosettes placed in cutting compost and put in a frame or propagator in either spring or late summer/ autumn.

A display of Porophylla saxifrage hybrids.

Saxifraga × *burnatii* ☼ ◑

This is one of the encrusted saxifrages and is a hybrid between *S. cochlearis* and *S. paniculata*. Its appearance is that of a tight cushion composed of lime-encrusted rosettes with some of the leaves 'sticking out' from the cushion. A large specimen would be 15cm (6in) across and with its wiry, arching sprays of small white flowers produced during the summer months makes a nice plant for a gritty sink garden or crevice between rocks on an alpine bed. Single rosettes, detached with a little bit of stem, can be treated as cuttings at any time of the year but best results are when taken from spring to autumn.

Saxifraga 'Cloth of Gold' ☼ ◑

It is not difficult to see how this plant got its name with spreading golden-yellow mounds of rosettes that can reach over 30cm (12in) in diameter with the added bonus of short-stemmed white flowers in spring. Any dryness should be avoided as this will often lead to unsightly scorched brown patches. Should this occur, pull out the burnt growth and fill the hole

Saxifraga 'Cloth of Gold'.

with a mixture of sand and compost; this will encourage new shoots to grow back. Suitable planting positions are therefore where there is plenty of moisture though still well drained, along with some sun to maintain a good bright golden colour. New plants can be made from several long shoots bunched together complete with their brown stems and planted directly into a sandy compost in early autumn. *Saxifraga* 'Cloth of Gold' is a member of a group known as 'mossy saxifrages' due to their moss-like appearance. Other members of this group have lush green or variegated foliage, white, pink or red flowers, and are easily grown in similar conditions to *S.* 'Cloth of Gold'.

Saxifraga cochlearis ☼ ◑

A deservedly popular species restricted in the wild mainly to the Maritime Alps where it inhabits mostly shaded vertical rocks on limestone. In the alpine garden limestone rocks are not necessary for success – most sharply drained soils are suitable – though being a small plant it can be lost on a large rock garden and for this reason raised beds, sink gardens and the alpine house are perhaps the best places for cultivation. The plant has a dome-like appearance formed of rosettes of tiny, hard, silvery-grey leaves, the whole rarely more than 12cm (5in) in diameter with pretty, red stemmed sprays of white flowers in summer.

Saxifraga cochlearis.

Saxifraga cotyledon 'Slack's Ruby Southside' AGM ☼ ◑ A

Saxifraga cotyledon 'Slack's Ruby Southside' AGM.

This encrusted saxifrage, when well grown, has a dense and distinctive 30cm (12in) pyramid of blood-red, white-edged flowers in early summer produced from 10cm (4in) wide bristly-edged green rosettes. An acid to neutral gritty and well-drained compost suits it best, avoiding hot dry sites. It grows well on all types of rock garden with sun and also makes a good show for an alpine house. Propagated by cuttings.

Saxifraga grisebachii AGM ☼ ◑

The full name is actually *Saxifraga federici-augusti* subsp. *grisebachii*; however, the species is well known to alpine gardeners as *grisebachii* and this fits on a label more easily! The grey-green rosettes can measure 7cm (3in) in diameter and form beautiful geometric

Saxifraga grisebachii AGM.

patterns, with colonies comprising several rosettes developing after a number of years. Flowering spikes, which last until May, begin to grow in late winter, slowly rising upwards not unlike a shepherd's crook, and are covered in short red-purple hairs and red leafy bracts. The flowers themselves are in rows along the stem, and are small and reddish. A large plant in flower is a truly magnificent sight. It requires perfectly drained soil, ideally containing plenty of sharp sand and some limestone chippings. In areas of high winter rainfall it is more safely grown under cover for the winter months – an alpine house would be the perfect place. It can be propagated from seed to produce a flowering plant in about three years, or individual rosettes can be taken as cuttings. It is native to northern Greece.

Saxifraga 'Hare Knoll Beauty' ☼ ◑

The majority of summer-flowering saxifrages have white flowers, but this variety bucks the trend with graceful sprays of mid-pink flowers. Neat and attractive grey-blue rosettes along with a hardy disposition and ease of culture make 'Hare Knoll Beauty' a good choice for summer colour, whether on a raised bed, rock garden or in a sink garden. Sun and gritty, well-drained soil provide the right conditions where colonies of rosettes about 20cm (8in) across can be expected. Cuttings are the best means of increase.

Saxifraga 'Hare Knoll Beauty' (and *Dactylorhiza* hybrid).

Saxifraga 'James' ☼ ◑

Saxifraga 'James'.

One of the Porophylla group of *Saxifraga* and among the best of the yellow varieties, with reddish buds on very short stems produced freely and early in the year at the beginning of spring. The dark green, glossy leaves are tiny and form small rosettes that grow together into 4cm (1.5in) high cushions reaching 15cm (6in) across in a well grown plant. Best cultivated in either a sink garden with other slow-growing and well-behaved inhabitants or in an alpine house where the flowers achieve perfection undamaged by ice or rain. Cuttings are best taken in spring after flowering or early autumn.

Saxifraga 'Jenkinsiae' AGM ☼ ◑

This splendid cultivar, raised around 1920, has stood the test of time with its large, rounded pale pinkish, dark-eyed flowers on reddish stems from very compact spiny leaved cushions. Very hardy, the flowers open undamaged even after the buds have been encased in ice, for this is another cultivar that flowers very early in spring. Suitable for both a

Saxifraga 'Jenkinsiae' AGM.

rock garden and containers where plants can span 20cm (8in) after 10 years.

Saxifraga longifolia ☼ ◑

Without doubt one of the very finest species – very beautiful in both flower and foliage – it is native to the Pyrenees where it grows on vertical cliffs throwing out a magnificent plume made up of hundreds of white flowers in summer. Unfortunately the plant is nearly always monocarpic – that is, it dies after the fabulous flowering, which produces thousands of seeds to propagate itself. The plant, however, does take several years to reach a size large enough to produce the flower spike and during those years has great beauty in the perfectly symmetrical rosette of long narrow grey-green pointed leaves achieving a diameter of up to 16cm (6in). Ideally it should be grown vertically between rocks or at least in very gritty soil and is lovely for an alpine house or cold frame. Propagate by seed.

Saxifraga oppositifolia 'Splendens' AGM ☼ ◑

Dense, prostrate mats of creeping shoots are clothed in small, bristly, dark-green leaves, studded with good sized rounded purplish-pink flowers in early spring. Mats up to 30cm (12in) across quickly grow if given well-drained gritty soil in a sunny position and plenty of moisture during the growing season. It enjoys growing over rocks – a habit that can be used to good effect when planted at the edge of a raised bed, so allowing shoots to cascade down. Shoots often root along their length particularly in autumn and if these are removed they can be used as cuttings to make new plants. A very hardy alpine found widely throughout Europe, North America and Northern Asia.

Saxifraga longifolia.

Saxifraga oppositifolia 'Splendens' AGM.

Saxifraga paniculata v. *minutifolia* ☼ ◑

Saxifraga paniculata has a number of varieties, all of them making good alpine garden plants on account of their attractiveness, hardiness and ease of cultivation. In this variety the species is at its tiniest with rosettes, often smaller than 1cm (0.5in) in diameter, of bristly glaucous leaves forming a very dense, flat cushion less than 1cm (0.5in) high. Flowers are small and white on branching wiry stems of up to 10cm (4in) produced during summer. This tiny form is perfect in a sink garden or container containing gritty compost and can be propagated by division or making cuttings of the rosettes. It is widely distributed in mountains of Europe and parts of North America. *S. paniculata* and the variety 'Rosea' are larger and suitable for planting in any well-drained, sunny spot in the alpine garden.

Saxifraga 'Polar Drift' ☼ ◑

Two outstanding features of this relatively new hybrid are the wonderful blue-grey, compact rosettes of silvery edged leaves and large, showy spike of white flowers produced in early summer. It makes a beautiful plant for any rock garden feature but is especially attractive growing among rocks on a slope where the flower spike can arch over. It also makes a splendid plant for a pot in an alpine house. Well-drained gritty soil is important along with some sun, which helps maintain the colour of the rosettes.

Saxifraga 'Sulphurea' ☼ ◑

A Porophylla hybrid of *S. burseriana* with very compact spiny grey-green rosettes and red-stemmed large pale yellow flowers, 3cm wide (1.5in). It is slow growing and ideal for sink gardens and raised beds.

Saxifraga 'Tumbling Waters' AGM ☼ ◑

Awarded an RHS 'Award of Garden Merit' in 1920 for its garden worthiness and with one of the most spectacular flowering spikes of the genus whose arching spray can approach 1m (40in) and comprise up to 1,000 individual flowers. Leaf rosettes grow up to 15cm (6in) in diameter and are of stiff grey narrow leaves with hooked tips. It likes a sunny and well-drained soil and is perfectly placed where the arching flower spike can hang down, for example at the edge of a raised bed or rocky outcrop. Small side-offsets can be removed and treated as cuttings during the growing season.

Scilla siberica AGM ☼ ◑

A well-known, easy and beautiful early spring flowering bulb from Siberia with abundant brilliant blue pendant bell-shaped flowers on 10–12cm (4–5in) stems. It grows in any decent soil, sun or part shade, and increases well by self-seeding to create drifts of blue.

Scleranthus biflorus ☼

For those who love tactile plants, *Scleranthus biflorus* is quite irresistible, for when grown in an open and sunny position it presses its dense cushion of tiny pointed leaves to the ground like a hard moss. The curious flowers are rather insignificant but are followed by slightly more interesting seeds, like minute yellowish candles sitting on the cushion. It grows best in gritty or sandy, well-drained soil in sun. The maximum size is around 15cm (6in), making it small enough for sink gardens and containers but equally happy on a rock garden.

Scleranthus biflorus.

Propagation is easiest from long stem cuttings taken in autumn. It is native to New Zealand and Tasmania.

Sedum 'Dragon's Blood' AGM ☼

A splendid foliage plant for any sunny and well-drained position, *Sedum* 'Dragon's Blood' looks especially pleasing trailing down a bank or over the edge of a wall. The prostrate stems are clothed in toothed oval leaves that can vary in colour according to conditions from deep red-brown to a bright mahogany. The leaves drop off for the winter months. Pink flowers appear during late summer and autumn in flat heads at the ends of the stems.

Sedum 'Dragon's Blood' AGM.

Sedum 'Lidakense' AGM ☼

A delightful *Sedum* from the moment it sprouts from the ground in spring with its wiry arching shoots clad in rounded fleshy grey leaves that turn to pinkish shades as summer wears on, culminating in flattish heads of small red-pink flowers that last into autumn. It is small enough and indeed ideal for a sink garden as well as any sunny rock garden feature, and is easily increased from cuttings taken in summer. Height attained is around 10cm (4in) with a spread of 20cm (8in).

Sedum 'Lidakense' AGM.

Sedum obtusatum ☼

Under suitable conditions the fleshy, rounded glossy leaves have an intense reddish colour and form a dense mat above which flat heads of yellow flowers appear during summer months. It grows best in full sun in very gritty soil, to a height of 5cm (2in) and 15cm (6in) spread. Native to North America.

Sedum obtusatum.

Sempervivum ☼

Sempervivum species and hybrids are justifiably popular due in much part to their enormous variation in leaf colour, which includes yellow, green, red, grey, and bronze. Sometimes leaves have two colours; sometimes they are covered in hairs, giving a cobweb-like appearance. All form their leaves into a rosette varying in size from 1cm (0.5in) to 12cm (5in), which eventually gives rise to an erect and sturdy stem with several star-like flowers – these are often a dull pink but can be bright pink, greenish-yellow or white. They like to be grown in very well-drained compost in full sun and lend themselves to shallow containers where they can be taken under cover for winter as many benefit from the protection from excess rain that this affords. Propagation is by simply detaching the numerous offsets and planting them straight into gritty compost. They can survive periods of drought and neglect, making them ideal subjects for introducing children to gardening. If growing sempervivums in containers on their own then use compost containing equal parts by volume of loam, grit and sharp sand. From the hundreds of different species and hybrids to choose from the following are really good ones for growing outdoors: *S. calcareum*, a tough species; *S. nevadense*, which has tight, rounded globe-like rosettes of yellowish-green; *S.* 'Bronze Pastel', which displays compact hummocks of copper-brown; *S.* 'Beta', one of the tougher red varieties; and *S.* 'Virgil', a wonderful olive-grey, fairly tough outdoors.

Sempervivum × barbulatum 'Hookeri' *and S.* 'Bronze Pastel'.

Sempervivum arachnoideum ☼ ∩

Affectionately known as the 'cobweb houseleek' due to the covering of hairs like a spider's web, this characteristic is reflected in the species name and is an adaptation for water conservation and insulation in their mountain habitat. There are a number of named forms differing in size of rosette: 'Laggeri' is one of the smallest with rosettes around 1cm (0.5in) diameter (the smaller ones are more tolerant of winter wet if grown outdoors); 'Clärchen' is one of the larger and is especially hairy, and is best grown under cover in wet districts. Its flowers are some of the best among sempervivums, being a good pink and sometimes near red.

Silene acaulis 'Frances' ☼ ◑

Silene acaulis is more commonly known as 'moss campion' for its moss-like appearance and pink flowers that appear in spring, continuing on and off throughout summer. 'Frances' is a golden foliage form found by chance growing wild and is a splendid and easily cultivated alpine for a not too dry position in good light, where the cushions can spread to 20cm (8in) or more while just a centimetre (0.5in) high. The flowers are pale pink and completely stemless, studding the cushion in late spring. It is well suited to sink gardens where it will mould itself around rockwork and over the edge like a golden carpet. It comes from mountainous areas in the northern hemisphere. Taken in spring, stem cuttings just turning brown at the base are used for propagation.

Silene acaulis 'Frances'.

Silene alpestris 'Flore Pleno' AGM ☼ ◑

Silene alpestris 'Flore Pleno' AGM.

Loose tufts of leafy rosettes are pleasant enough but easily eclipsed by the abundant semi-double, gleaming white flowers on dark wiry branching stems during the summer. 'Flore Pleno' grows well in any good soil and sunny position, making a showy mound on rock gardens and raised beds. Deciduous in habit, the newly emerging shoots in spring can be cut off and made into cuttings to produce new plants for the following year. It is a British native plant.

Sisyrinchium 'Californian Skies' ☼

Sisyrinchium foliage is grass- or iris-like in appearance, growing in tufts of fans from little creeping rhizomes. 'Californian Skies' has flowers typical of the genus, held at the tips of 10cm (4in) stems during midsummer. In this case they are larger than most and a good blue-mauve, and unlike many of the species it does not seed about so as to become a nuisance. A sunny site and well-drained soil are all it needs along with division every few years and replanting in new soil to maintain vigour. All members of the genus have their origins in North America.

Soldanella carpatica ◑ ▲

A favourite photographic subject for anything promoting the Alps of Europe! This charming genus contains a handful of species flowering in early spring and with loose clusters of beautiful fringed mauve or purple bells. *Soldanella carpatica* is easily identifiable even when not in flower by having a dark purple reverse to the rounded deep green leaves. It thankfully seems to be more reliable in

Soldanella carpatica.

flowering than other species, for the flowers are an absolute delight. Gritty compost and the avoidance of hot, dry conditions, with plenty of moisture, sited perhaps in a sink garden or lightly shaded spot on a raised bed, should ensure success. The height in flower is around 12cm (5in) with a spread of 15cm (6in). Established clumps can be propagated by careful division after flowering.

Sorbus reducta AGM ☼ ◑

A truly miniature 'tree', small enough even for a container where its sturdy little branches with glossy pinnate leaves add height and character. Flat heads of off-white flowers in summer result in quite large pink-red berries from which seeds can be extracted to raise new plants. The foliage takes on good red autumn colours too. Maximum size after several years is around 30cm (12in). It is suitable for any rock garden feature in sun or part shade.

Thalictrum kiusianum ◑ ●

This miniature member of a genus better known for its herbaceous species is an

Thalictrum kiusianum.

absolute treasure with mildly creeping rhizomes that develop 10cm (4in) mounds of tangled stems bearing dainty foliage throughout summer. The tassels of mauve-pink flowers are produced over a long period from June until autumn on twiggy stems up to 20cm (8in) tall. It will be unhappy with any dryness at the roots, ideally preferring cool, humus-enriched soil, perhaps in the shade of a rock. An established clump can be lifted for propagation during dormancy and the mass of little rhizomes divided into smaller clumps. It is a native of Japan.

Thymus ☼

The numerous forms of carpeting thymes are among the best alpines for establishing among paving stones in sunny, well-drained soil, where they self--sow to create a patchwork of prostrate mats covered in summer with either reddish-purple, pink, lilac or white flowers, much frequented by insects for nectar and pollen alike. *T.* Coccineus Group is a good near-red variety. Self-sown plants often seem to do better than purposely planted ones especially when trying to establish them among paving. Seedlings also flower at slightly different times and so help to prolong the season of colour. For propagation of named cultivars, cuttings must be used and these are best taken in spring or early summer.

Thymus serpyllum.

Trillium chloropetalum AGM ◑ ●

Trillium is a genus of mainly North American woodland plants with a handful of species occurring in Asia. A shared characteristic among them is that various parts of the plant are divided into three – most notably in the leaves, which are composed of three leaflets, and the flowers, which have three petals. The unfurling leaves of *Trillium chloropetalum* are very attractive with purplish-brown flecks and blotches over their dark green surfaces. Under suitable conditions they will make bold patches of ground cover from large slowly clump-forming rhizomatous roots. Opening leaves reveal the curious flowers standing upright at the centre of the three leaflets; the colour varies but normally is a dark brownish-purple and about 6cm (2.5in) tall. The whole plant grows to about 25cm (10in). Plant in a

Trillium chloropetalum AGM.

position of light shade and in good compost, or grow in a deep container or pot, again in part shade. Division is best carried out on large clumps after flowering, or all species can be raised from seed sown as soon as ripe – but be warned: you will have to wait for five years or more before the first flower appears!

Trillium pusillum var. *pusillum* ◑ ● ▲

A very pretty species, rather delicate in appearance but completely hardy, although its small size makes it best suited to a raised bed or container planted along with other treasures that appreciate that little bit of extra care. The pale pink flowers open like 'normal' flowers and are held at a slight angle on stems of about 10cm (4in); the neat foliage is mid-green and unmarked. A well-drained, moisture-retentive compost containing leaf mould gives good results. Propagation is by careful division or seed.

Trillium pusillum var. *pusillum*.

Trollius pumilus ☼

Native to the Himalayas, the beautiful bronze-green flower buds open to sculptural yellow buttercups up to 3cm (1.5in) across in early summer. The neat foliage grows in tufts usually no more than 10cm (4in) in height, and the flowers are held just above. It is very hardy and tolerant of most conditions except drought and deep shade. Abundant seed is produced, and this is the best means of increase.

Tulipa tarda AGM ☼

A very rewarding tulip from Central Asia with up to five star-shaped creamy-white tipped, golden yellow flowers on 5–10cm (2–4in) stems, perfectly set off by the pleasing foliage. They are eye-catching and perfect for all manner of sunny rock garden features containing gritty soil where, with luck, it may self-sow.

Tulipa tarda AGM.

Veronica prostrata AGM ☼

A very easy rock plant that can succeed where others may fail. It looks at its best hanging over rockwork or a wall top where the freely produced spikes of small, bright blue flowers can be admired close up. Prostrate stems up to 20cm (8in) in length are clothed in small, toothed leaves. Best results are achieved in a well-drained, sunny position. It is sometimes liable to becoming rather straggly – a condition easily remedied by giving it a hard clip over after which vigorous new growth will soon restore a healthy specimen. Soft green cuttings root well in a propagator during the growing season.

Viola cornuta 'Alba Minor' ◑

Viola cornuta is a hardy perennial *Viola* from the Pyrenees and the Picos de Europa in northern Spain. An easy plant for the garden, where it makes clumps of fresh green leaves and flowers from late spring until autumn, with abundant mauve-blue flowers reaching 20–25cm (8–10in) in height by 40cm (16in) spread. 'Alba Minor' is a dwarf white-flowered form more suitable for the alpine garden with a maximum height of just 10cm (4in). Easily grown in sun or shade and propagated simply by division or cuttings during the growing season, they respond well to cutting back hard with vigorous new growth and prolonged flowering.

Vitaliana primuliflora ☼

Beautiful in and out of flower, this very hardy and long-lived alpine deserves to be in every alpine garden. It forms a compact evergreen cushion 2–3cm (1–1.5in) high and around 15cm (6in) across made up of numerous stems clothed in silvery grey-green needle-like leaves and beautiful, stemless clear yellow flowers in spring. It looks great planted alongside the brilliant blue stars of *Gentiana verna*. Give it a well-drained and bright position.

Vitaliana primuliflora.

Zaluzianskya 'Orange Eye' ☼ A

Of comparatively recent introduction, this striking alpine has quickly become popular with its orange-centred, white flowers and pervasive sweet fragrance. It grows into a low hummock made up of aromatic, dark-green, toothed leaves with flowers held just above in late spring and early summer, sometimes with a more modest flowering later in the year. It can certainly withstand temperatures as low as –15°C (5°F) but benefits from having protection from cold, drying winds. Plenty of moisture is required in the growing season in an acid to neutral soil. Height is approximately 10cm (4in) with a spread of 25cm (10in). Soft green shoots taken in summer make the best cuttings.

Zaluzianskya 'Orange Eye'.

Zaluzianskya ovata.

Zaluzianskya ovata ☼ A

Similar to the preceding *Zaluzianskya* but larger in all its parts, *Z. ovata* has pretty white, red-backed flowers, and enjoys similar cultural conditions.

Zauschneria californica 'Dublin' AGM ☼

A native of California, the brilliant orange-red tubular-shaped flowers are freely produced on a low spreading bush of small narrow grey-green leaves. This is a definite sun-lover requiring well-drained soil and is very useful for flowering late in the season – often still in full flower when cut down by the first frosts of autumn. Unfortunately it can be killed in severely cold and wet winters if temperatures dip below around –12°C (10.4°F) so it is worthwhile rooting a few cuttings during the summer and keeping them protected just in case. A good plant for a raised bed or wall top where the low mounds can hang over in a mass of stunning flowers for several weeks from midsummer onwards. It can grow to 25cm (10in) in height, spreading to 40cm (16in) or more.

Zauschneria californica 'Dublin' AGM.

Zauschneria 'Pumilio' ☼

This is a hardier plant than *Z. californica* but with very similar red-orange flowers in summer. The whole plant is completely prostrate in habit, forming a dense mat barely above 5cm (2in) with a spread of 30cm (12in). The foliage is covered in hairs giving a greyish, woolly appearance. It can be propagated from the small green shoots that come up at the perimeter of the plant in spring.

Zauschneria 'Pumilio'.

TABLE OF FLOWERING TIMES

This list is intended as a guide as to when plants are in flower or of interest – climatic variations will of course influence precisely when a plant begins to flower and for how long. (Flowering times are based on those personally observed at 900ft altitude in the UK.) Where a plant is denoted as flowering in two periods this generally means that flowering covers the latter part of one period and the early part of the next. Where, for example, a plant has both 'spring' and 'year round', the flowers appear in spring but the foliage or form is also of year-round interest.

NB: Spring is March, April and May; Summer is June, July and August; Autumn is September, October and November, and Winter is December, January and February.

PLANT NAME	SPRING	SUMMER	AUTUMN	WINTER	YEAR ROUND
Aethionema 'Warley Rose' AGM		✔			
Allium amabile		✔			
Allium flavum AGM		✔			
Allium sikkimense		✔			
Anacyclus pyrethrum var. *depressus*	✔	✔			
Androsace carnea subsp. *brigantiaca*	✔				
Androsace carnea subsp. *laggeri* AGM	✔				
Anemone nemorosa	✔				
Anemone trullifolia	✔	✔			
Antennaria dioica	✔	✔			
Anthyllis vulneraria var. *coccinea*	✔	✔			
Aquilegia bertolonii AGM	✔	✔			
Arenaria balearica	✔				

PLANT NAME	SPRING	SUMMER	AUTUMN	WINTER	YEAR ROUND
Arisaema sikokianum	✔				
Armeria juniperifolia AGM	✔				
Aster coloradoensis		✔			
Astilbe glaberrima		✔	✔		
Aubrieta	✔	✔			
Calceolaria tenella	✔	✔	✔		
Calceolaria uniflora var. *darwinii*	✔	✔			
Campanula betulifolia AGM		✔			
Campanula garganica 'Dickson's Gold'					✔
Campanula pulla		✔			
Campanula raineri AGM		✔			
Cardamine pratensis 'Flore Pleno'	✔				
Celmisia ramulosa var. *tuberculata*		✔			✔
Celmisia semicordata 'Slack Top Hybrids'		✔			✔
Centaurium scilloides	✔	✔	✔		
Chionohebe pulvinaris					✔
Chrysosplenium davidianum	✔				
Colchicum agrippinum AGM			✔		

PLANT NAME	SPRING	SUMMER	AUTUMN	WINTER	YEAR ROUND
Corydalis 'Kingfisher'	✔	✔	✔		
Cotoneaster microphyllus var. *cochleatus*					✔
Crepis incana AGM		✔			
Crocus	✔		✔		
Cyananthus microphyllus AGM		✔	✔		
Cyclamen coum AGM	✔			✔	
Cyclamen hederifolium AGM	✔		✔	✔	
Daphne cneorum		✔			
Daphne retusa	✔				✔
Delosperma basuticum	✔	✔			
Delosperma from Graaf Reinet	✔	✔			
Dianthus alpinus AGM	✔	✔	✔		
Dianthus microlepis	✔	✔			
Draba rigida var. *imbricata*	✔				
Dryas octopetala AGM	✔				
Dwarf conifers					✔
Edraianthus dinaricus		✔			
Epilobium glabellum		✔	✔		

PLANT NAME	SPRING	SUMMER	AUTUMN	WINTER	YEAR ROUND
Erigeron aureus 'Canary Bird' AGM	✔	✔	✔		
Erigeron karvinskianus AGM		✔	✔		
Erinus alpinus AGM	✔	✔			
Erodium × kolbianum		✔	✔		
Erodium × variabile 'Roseum' AGM		✔			
Erythronium dens-canis AGM	✔				
Erythronium 'Pagoda' AGM	✔				
Ewartia planchonii					✔
Fritillaria camschatcensis	✔				
Galanthus 'S. Arnott' AGM	✔				
Gentiana acaulis AGM	✔				
Gentiana autumn-flowering types			✔		
Gentiana paradoxa		✔	✔		
Gentiana saxosa		✔	✔		
Gentiana septemfida AGM		✔			
Gentiana verna	✔				
Geranium 'Apple Blossom'		✔			
Geranium 'Ballerina' AGM		✔			

PLANT NAME	SPRING	SUMMER	AUTUMN	WINTER	YEAR ROUND
Geranium dalmaticum 'Bridal Bouquet'		✔			
Geranium farreri		✔			
Geranium subcaulescens 'Splendens' AGM		✔			
Geum montanum AGM	✔	✔			
Gladiolus flanaganii		✔	✔		
Glaucidium palmatum AGM	✔				
Globularia meridionalis 'Hort's Variety'		✔			
Haberlea rhodopensis AGM	✔				
Hacquetia epipactis AGM	✔				
Helianthemum		✔			
Helichrysum milfordiae AGM	✔	✔			
Hepatica	✔				
Hosta	✔	✔	✔		
Iris reticulata AGM	✔			✔	
Jeffersonia dubia	✔				
Leptinella dendyi	✔				
Leucogenes leontopodium		✔			✔
Lewisia columbiana 'Alba'		✔			

PLANT NAME	SPRING	SUMMER	AUTUMN	WINTER	YEAR ROUND
Lewisia cotyledon AGM	✔	✔			
Lilium duchartrei		✔			
Linaria alpina		✔	✔		
Linum capitatum		✔			
Lithodora diffusa 'Picos'	✔	✔			
Meconopsis 'Lingholm'	✔				
Meconopsis quintuplinervia AGM	✔	✔			
Morisia monanthos 'Fred Hemingway'	✔	✔			
Narcissus	✔				
Olsynium douglasii AGM	✔				
Origanum amanum AGM		✔			
Oxalis enneaphylla 'Patagonia'	✔	✔			
Oxalis enneaphylla 'Sheffield Swan'	✔	✔	✔		
Oxalis laciniata	✔	✔			
Oxalis 'Slack Top Hybrids'	✔	✔			
Penstemon rupicola AGM	✔	✔			
Phlox	✔	✔			
Physoplexis comosa AGM		✔			

PLANT NAME	SPRING	SUMMER	AUTUMN	WINTER	YEAR ROUND
Phyteuma hemisphaericum		✔			
Polygala calcarea 'Lillet' AGM	✔	✔	✔		
Polygala chamaebuxus AGM	✔	✔			
Polygala chamaebuxus 'Rhodoptera'	✔	✔			
Potentilla eriocarpa		✔	✔		
Potentilla nitida 'Rubra'		✔			
Primula auricula AGM	✔				
Primula auricula 'Blairside Yellow'	✔				
Primula auricula 'Slack Top Red'	✔				
Primula denticulata 'Birthday Beauty'	✔				
Primula marginata AGM	✔	✔	✔		
Primula × *pubescens* 'Freedom'	✔				
Primula 'Slack Top Violet'	✔				
Primula scotica	✔	✔			
Primula sieboldii AGM	✔	✔			
Primula vulgaris AGM	✔			✔	
Pritzelago alpina	✔	✔			
Pulsatilla AGM	✔				

PLANT NAME	SPRING	SUMMER	AUTUMN	WINTER	YEAR ROUND
Ramonda myconi AGM	✔	✔			
Ranunculus alpestris	✔	✔			
Rhodiola rosea	✔	✔			
Rhododendron	✔				
Rhodohypoxis		✔	✔		
Roscoea cautleoides AGM		✔			
Salix 'Boydii' AGM					✔
Salix reticulata AGM					✔
Sanguinaria canadensis 'Plena' AGM	✔				
Saxifraga × *burnatii*		✔			✔
Saxifraga 'Cloth of Gold'		✔			✔
Saxifraga cochlearis		✔			✔
Saxifraga cotyledon 'Slack's Ruby Southside' AGM		✔			
Saxifraga grisebachii AGM	✔				✔
Saxifraga 'Hare Knoll Beauty'		✔			✔
Saxifraga 'James'	✔				
Saxifraga 'Jenkinsiae' AGM	✔				
Saxifraga longifolia		✔			✔

PLANT NAME	SPRING	SUMMER	AUTUMN	WINTER	YEAR ROUND
Saxifraga oppositifolia 'Splendens' AGM	✔				
Saxifraga paniculata var. *minutifolia*		✔			✔
Saxifraga 'Polar Drift'		✔			✔
Saxifraga 'Sulphurea'	✔				
Saxifraga 'Tumbling Waters' AGM		✔			
Scilla siberica AGM	✔				
Scleranthus biflorus					✔
Sedum 'Dragon's Blood' AGM		✔			
Sedum 'Lidakense' AGM		✔	✔		
Sedum obtusatum		✔			✔
Sempervivum		✔			✔
Silene acaulis 'Frances'	✔	✔	✔		
Silene alpestris 'Flore Pleno' AGM		✔			
Sisyrinchium 'Californian Skies'		✔			
Soldanella carpatica	✔				
Sorbus reducta AGM	✔	✔	✔		
Thalictrum kiusianum		✔	✔		
Thymus		✔			

PLANT NAME	SPRING	SUMMER	AUTUMN	WINTER	YEAR ROUND
Trillium	✔				
Trollius pumilus	✔				
Tulipa tarda AGM	✔				
Veronica prostrata AGM		✔			
Viola cornuta		✔	✔		
Vitaliana primuliflora	✔				
Zaluzianskya 'Orange Eye'	✔	✔			
Zaluzianskya ovata	✔	✔			
Zauschneria californica 'Dublin' AGM		✔			
Zauschneria 'Pumilio'		✔			

Glossary

This glossary gives the meanings of both botanical and horticultural terms used within this book.

Acid Soil with a pH of less than 7. Used for ericaceous plants.
Alpine A plant growing above the tree line in mountainous regions, but may in practice be any small hardy plant suitable for cultivating in a rock garden.
AGM Royal Horticultural Society's Award of Garden Merit.
Aspect The position of a plant, feature or object relative to the sun or points of the compass.
Boss A large stud or blob of stamens.
Bract Modified leaf-like structure enclosing or combined with a flower or cluster of flowers. Can be small or large, sometimes coloured and often gives some protection to the flower.
Bulbous In plants, a swollen storage organ from which arise foliage, flowers and roots.
Bullate Having a surface of indentations and/or ridges with regard to leaves.
Compost A mixture of ingredients that could for example include grit, sharp sand, leaf-mould, garden soil, garden compost, etc.
Corymb Flattish flower head composed of several flowers.
Crowbar A length of round iron bar, bluntly pointed or flattened at one end and used as a lever for moving large rocks, etc.
Deciduous A plant that sheds its leaves annually, and re-grows them the following year.
Decking Timber planks used in the construction of outdoor garden features such as patios. Usually treated for protection against moisture.
Desiccant Material for absorbing moisture.
Dibber Bluntly pointed tool used to make a hole in soil for the insertion of cuttings or seedlings.
Ericaceous Denotes plants of the heather family which require an acid soil or a soil with a pH of less than 7.
Farina Powdery dusting naturally found on the flowers and foliage of some plants.
Form Used to describe variation within a species, e.g. a large-flowered form.
Garden compost Decayed garden material usually made on a compost heap.
Genus Term of plant classification between family and species.
Hardy A plant able to survive frosts.
Humus Decomposed organic vegetable matter, composed of leaves, compost, etc.
Inflorescence Parts composing a flower head including flowers, stalks, bracts and stems.
Lamina The surface of a leaf, not including the stalk.
Leaflets A small leaf, or component of a compound leaf.
Liquid feed Water containing dissolved plant food.
Monocarpic A plant that flowers once in its lifetime and then dies.
Node The point from which side shoots arise or where leaf stalks are attached.
Outcrop Rock formations above ground.
Palmate Refers to a leaf shape that is hand-like or with lobes.
Petiole Stalk attaching the leaf blade or lamina to the stem.
pH A numerical measure of acidity or alkalinity. Below pH7 soil increases in acidity, above pH7 it becomes increasingly alkaline or limey.
Pinnate Referring to leaf shape, with leaflets on either side of a stem like a feather.
Scree Sloping formation of rocks, stones and chippings caused by cliff rocks being shattered by frost. Simulated in domestic rock gardens.
Soil Upper layer of earth in which plants grow, typically consisting of broken down organic matter, clay and particles of rock.
Spathe A sheathing bract that encloses and partially protects the flower cluster.
Species Term of plant classification below genus, which contains similar plants.

Sport A genetic mutation resulting in a change to a particular plant characteristic, for example flower colour or leaf size.

Sterilized loam Soil that has undergone treatment either from chemicals or by heating to a temperature of 82–93°C (180–200°F) for ten minutes. This is primarily to destroy weed seeds and the roots of weeds. Harmful as well as harmless organisms are also killed.

Stolon A shoot, normally horizontal, which roots at the tip to develop a new plant.

Stratification Process of seed treatment involving warm or cold periods to aid germination.

Subsoil Section of earth below the topsoil.

Subspecies Further classification of plant below species, often abbreviated to subsp. e.g. *Saxifraga oppositifolia* subsp. *oppositifolia.*

Sump An excavated hole in the ground filled with drainage material to enable excess water to drain away.

Tesselated Having a chequered pattern on the petals.

Topsoil The top layer of soil.

Translocation Movement of fluids around a plant's internal vascular system.

Turgid In plants, a term used to describe leaves or stems that contain their required amount of water.

Turks-cap Common name for *Lilium martagon*, which has flowers vaguely resembling a turban.

Umbel Botanical term for the arrangement of flowers in a flower head, where the flowers originate from one point, like the spokes of an umbrella.

Viable Able to germinate.

Alpine nurseries and seed suppliers

In all cases, website addresses are shown where available, otherwise a postal address is given.

ALPINE NURSERIES

The following section contains a selection of good 'grower' nurseries that propagate most, if not all, of their plants and also open regularly to the public. They are known from personal experience or through one of the specialist alpine societies' lists. Specialist nurseries take great pride and care in the production of their plants by using good-quality composts that result in sturdy plants ready for the garden.

A good nursery differs from a general garden centre in that the nursery will have propagated and grown the plants that are offered for sale, as well as stocking many of the unusual species and varieties that enthusiasts are looking for. Unlike most garden centres, specialist growers can give expert advice based on personal experience. For in-depth advice it is often worth phoning ahead to agree a mutually convenient time, as small working nurseries (usually with only one or two staff) are invariably trying to juggle several tasks. Whilst the nursery may be open to the public and simple advice is freely offered, behind the scenes an emptied tray of seedlings may desperately need attention on the potting bench, or a van may need to be loaded up for a 5.00am journey to a show the following day.

UK

Slack Top Alpine Nurseries (owned and run by the author), 1 Waterloo House, 24 Slack Top, Hebden Bridge, West Yorkshire HX7 7HA. Tel: 01422 845348.
www.slacktopnurseries.co.uk

Aberconwy Nurseries, Graig, Glan Conwy, North Wales LL28 5TL.
Tel: 01492 580875.

Ardfearn Nurseries, Bunchrew, Inverness, Scotland IV3 8RH.
Tel: 01463 243250.
www.ardfearn-nursery.co.uk

Edrom Nurseries, Coldingham, Eyemouth, Berwickshire TD14 5TZ.
Tel: 01890 771386.
www.edrom-nurseries.co.uk

Glendoick Gardens (specializes in dwarf rhododendrons), Glencarse, Perth, Scotland PH2 7NS. Tel: 01738 860205.
www.glendoick.com

Hartside Nurseries, Alston, Cumbria CA9 3BL.
Tel: 01434 381372.
www.plantswithaltitude.co.uk

Pottertons Nurseries, Moortown Road, Nettleton, Caistor, Lincolnshire LN7 6HX.
Tel: 01472 851714.
www.pottertons.co.uk

Tile Barn Nursery (specializes in cyclamen), Standen Street, Iden Green, Benenden, Kent TN17 4LB.
Tel: 01580 240221.
www.tilebarn-cyclamen.co.uk

Timpany Nurseries, 77 Magheratimpany Road, Ballynahinch, Co. Down, N. Ireland BT24 8PA.
Tel: 028 9756 2812.
www.timpanynurseries.com

Europe

Jakob Eschmann, Waltwil, 6032 Emmen, Switzerland.

V D Beuken Alpines, Zegersstraat 7, 5961 XR, Horst (L) Netherlands.

USA

Mt Tahoma Nursery, 28111, 112th Ave E, Graham, Washington 98338 USA.
www.backyardgardener.com/mttahoma

Siskiyou Rare Plant Nursery, 2825 Cummings Road, Medford, Oregon 97501 USA.
www.siskiyourareplantnursery.com

SEED SUPPLIERS AND EXCHANGES

Alpine Garden Society (seed exchange).
www.alpinegardensociety.net

Jelitto Staudensamen GmbH, Postfach 1264, 29685 Schwarmstedt, Germany. www.jelitto.com

Mojmir Pavelka, Euroseeds, Czechoslovakia. www.pavelkaalpines.cz

New Zealand Alpine Garden Society (seed exchange), NZAGS, PO Box 2984, Christchurch, New Zealand.

North American Rock Garden Society (seed exchange). www.nargs.org

Scottish Rock Garden Club (seed exchange). www.srgc.org.uk

Silverhill Seeds (South African seeds), Cape Town, South Africa. www.silverhillseeds.co.za

Alpine gardens and shows

Alpine gardens and displays

UK

Alpine Garden at Slack Top,
1 Waterloo House, 24 Slack Top,
Hebden Bridge, West Yorkshire
HX7 7HA.
www.slacktopnurseries.co.uk

Alpine Garden Society Centre
and Garden, Avon Bank,
Pershore, Worcestershire
WR10 3JP.
www.alpinegardensociety.net

Branklyn Garden,
116 Dundee Road, Perth,
Scotland PH2 7BB.
www.nts.org.uk/Property/12/

Parcevall Hall, Skyreholme,
Skipton, N. Yorkshire
BD23 6DE.
www.parcevallhallgardens.co.uk

RHS Garden Harlow Carr, Crag
Lane, Harrogate, N. Yorkshire
HG3 1QB.
www.rhs.org.uk/harlowcarr

RHS Garden Wisley, Woking,
Surrey GU23 6QB.
www.rhs.org.uk/wisley

Royal Botanic Garden, Inverleith
Row, Edinburgh, Scotland
EH3 5LR.
www.rbge.org.uk

Europe

Arctic-alpine Botanic Garden,
University of Tromsø, NO-9037
Tromsø, Norway.

Brno Botanical Garden, Mendel
University, Brno, Czech Republic.

Chanousia Alpine Botanical
Garden, La Thuile, Valle d'Aosta,
Italy.
www.chanousia.org

Gothenburg Botanical Garden,
Gothenburg, Sweden.
http://gotbot.se/kulturvast_templates/Kultur_ArticlePageWide.aspx?id=56241

Lautaret Alpine Botanical
Garden, Villar d'Arène, Hautes-Alpes, France.
http://sajf.ujf-grenoble.fr/

Patscherkofel Alpine Garden,
Patscherkofel 7, 6082 Patsch,
Innsbruck, Austria.
www.uibk.ac.at/bot-garden/alpen/eindex.html

USA

Betty Ford Alpine Gardens,
183 Gore Creek Dr Ste. 7, Vail,
Colorado 81657, USA. www.bettyfordalpinegardens.org

Denver Botanical Gardens, 1007
York Street, Denver, Colorado
80206, USA.
www.botanicgardens.org

FLOWER SHOWS

The huge number of flower shows and plant fairs taking place across the UK each year are an excellent way to source the best in alpine plants, direct from the grower. Apart from being a great day out it can be a good opportunity to ask advice, share knowledge and simply enjoy the vast array of alpines that are usually on display. The list below details some of the larger national events and there are also many smaller specialist plant fair events and Alpine Garden Society group events across the country all year.

UK

RHS Spring Flower Shows,
Cardiff (April).

Harrogate Spring and Autumn
Flower Shows (April and
September).

Malvern Spring and Autumn
Shows (May and September).

RHS Chelsea Flower Show (May).

Gardening Scotland (June).

BBC Gardeners' World Live,
Birmingham (June).

Hampton Court Palace Flower
Show (July).

RHS Flower Show at Tatton Park
(July).

Europe

Courson, France (May).
Chaumont-sur-Loire, France
(April–October).

USA and Canada

Philadelphia International Flower
Show (March).

Australia and New Zealand

Ellerslie International Flower
Show, New Zealand (March).

International societies and special interest groups

Joining one of these societies can often be the best way to learn more about alpines and take part in events. Most of them also offer an annual seed exchange.

UK

Alpine Garden Society, AGS Centre, Avon Bank, Pershore, Worcestershire WR10 3JP.
www.alpinegardensociety.net

Scottish Rock Garden Club, PO Box 14063, Edinburgh EH10 4YE.
www.srgc.org.uk

The Hardy Plant Society, Little Orchard, Great Comberton, Pershore WR10 3DP.
www.hardy-plant.org.uk

USA

North American Rock Garden Society, PO Box 18604, Raleigh, NC, 27619-8604, USA.
www.nargs.org

New Zealand

New Zealand Alpine Garden Society, GW Clark, 47 Every Street, Anderson's Bay, New Zealand.
www.nzags.com

Further reading

There are many excellent books about alpines and those recommended below are written by experts in their fields.

BOOKS

Blanchard, J.W., *Narcissus: A Guide to Wild Daffodils* (The Alpine Garden Society, 1990)

Case, F.W. Jr. and R.B. Case, *Trilliums* (Timber Press, 1997)

Cobb, J.L.S., *Meconopsis* (published in association with The Hardy Plant Society by Timber Press, 1989)

Cox, P.A., *The Smaller Rhododendrons* (Batsford, 1985)

Good, J. and D. Millward, *Alpine Plants: Ecology for Gardeners* (Batsford, 2007)

Grey-Wilson, C., *The Genus Cyclamen* (Timber Press, 1988)

Griffith, A.N., *Collins Guide to Alpines* (Chancellor Press, 1964)

Halliwell, B., *The Propagation of Alpine Plants and Dwarf Bulbs* (Batsford, 1992)

Hills, L.D., *The Propagation of Alpines* (Faber & Faber, 1950)

Jermyn, J., *Alpine Plants of Europe: A Gardener's Guide* (Timber Press, 2005)

Leeds, R., *Alpine Bulbs in Containers* (Timber Press, 2005)

Mathew, B., *Dwarf Bulbs* (The Garden Book Club, 1973)

Mineo, B., *Rock Garden Plants: A Colour Encyclopaedia* (Timber Press, 1999)

Nichols, G., *Alpine Plants of North America* (Timber Press, 2002)

The Royal Horticultural Society, Plant Finder (Dorling Kindersley Ltd, 2010)

Salmon, J.T., *A Field Guide to the Alpine Plants of New Zealand* (Godwit, 1992)

Smith, G.F., B. Burrow, and D.B. *Low, Primulas of Europe and America* (The Alpine Garden Society, 1984)

Thomas, G.S., *The Rock Garden and its Plants* (J.M. Dent & Sons, 1989)

Webb, D.A., and R.J. Gornall, *Saxifragas of Europe* (Christopher Helm, 1989)

JOURNALS

These specialist societies publish regular journals and to receive a copy you usually need to join the relevant society. They are all printed in English.

Alpine Garden Club of British Columbia

Alpine Garden Society (AGS)
New Zealand Alpine Garden Society (NZAGS)

North American Rock Garden Society (NARGS)

Scottish Rock Garden Society (SRGC)

Acknowledgements

Thank goodness for midnight oil... and for my wife, Allison. There have been many long days and nights trying to keep up with the writing and photography schedule whilst juggling the day-to-day running of a full-time nursery and the flower show season, but luckily we both relish a challenge.

Thanks also go to the following: my sister Sally Stretch for her conscientious proofreading. Without her this book may have had a slight overuse of the comma; David Stretch and son Joel for their photographic assistance and slide scanning, even on Christmas Day; Ron Mitchell, my father, for his steady hand on the line illustration of rock garden construction; Andrew and Jenifer Wallis for their enduring support and professional expertise with imagery – particularly anything white; Val and Bob Smith for their supportive encouragement, and shared love of cake; Paula and Terry Mitchell for use of their immaculate alpine house and Jeni and Bob Wetton for their lovely green roof; our wonderful volunteers Eileen Smith and Elena Giacometti, for ensuring that seedlings and cuttings were potted up whilst we were busy with book photos, and for the use of Eileen's beautiful sempervivums and erigeron for photography.

Finally my love and thanks go to my wife Allison for her help with the research, copy editing and photography for this book, and most importantly for keeping me motivated through the challenges of competing book and nursery demands. Without her this book simply wouldn't have been possible.

Index

Other Gardening Books from Crowood

Blackburne-Maze, Peter *The Complete Guide to Vegetable Growing*
Clark, Emma *The Art of the Islamic Garden*
Cliff, Ann *The Value of Weeds*
Cooke, Ian *Designing Small Gardens*
Cooke, Ian *Exotic Gardening*
Cox, Freda *Garden Styles*
Cunningham, Sally *Ecological Gardening*
Dorey, Paul *Auriculas – an essential guide*
Ford, Richard *Hostas – an essential guide*
Gooch, Ruth and Jonathan *Clematis – an essential guide*
Gray, Linda *Herb Gardening*
Gregson, Sally *Ornamental Vegetable Gardening*
Gregson, Sally *Practical Propagation*
Hart, Simon *Tomatoes – a gardener's guide*
Hodge, Geoff *Pruning*
Jones, Peter *Gardening on Clay*
Larter, Jack *Tuberous Begonias*
Lavelle, Michael *Sustainable Gardening*
Littlewood, Michael *The Organic Gardener's Handbook*
Marder, John *Water-Efficient Gardening*
Nottridge, Rhoda *Wildlife Gardening*
Parsons, Roger *Sweet Peas – an essential guide*
Saunders, Bridgette *Allotment Gardening*